Water Governance - Concepts, Methods, and Practice

Series editors

Claudia Pahl-Wostl
University of Osnabrück
Osnabrück, Germany

Joyeeta Gupta
Fac. Earth & Life Sciences
University of Amsterdam
Amsterdam, The Netherlands

This book series aims at providing a platform for developing an integrated perspective on major advances in the field of water governance. Contributions will build bridges across established fields of expertise, across disciplinary perspectives, across levels from global to local and across geographical regions. Topics to be covered include conceptual and methodological advances capturing the complexity of water governance systems, institutional settings, actor constellations, diagnostic approaches, and comparative studies of governance systems. The book series encompasses monographs, textbooks and edited, coordinated volumes addressing one theme from different perspectives. The book series will address mainly a wider scientific audience, but will also provide valuable knowledge to interested practitioners. The sustainable management of fresh water resources and the ecosystem services that they provide is one of the key challenges of the 21st century. Many water related problems can be attributed to governance failure at multiple levels of governance rather than to the resource base itself. At the same time our knowledge on water governance systems and conditions for success of water governance reform is still quite limited. The notion of water governance aims at capturing the complexity of processes that determine the delivery of water related services for societal needs and that provide the context within which water management operates. Water governance is a fast growing field of scholarly expertise which has largely developed over the past decade. The number of publications in peer reviewed journals has increased from less than 20 in the year 2000 to nearly 400 in the year 2013. The increasing popularity of the term in science and policy has not lead to conceptual convergence but rather to an increasing vagueness and competing interpretations of how the concept should be understood, studied, and analyzed; which disciplines are involved; and what methodological approaches are most suitable for the study and analysis of water governance. The time seems to be ripe for comprehensive synthesis and integration.

More information about this series at http://www.springer.com/series/13400

Margot A. Hurlbert

Adaptive Governance of Disaster

Drought and Flood in Rural Areas

Springer

Margot A. Hurlbert
Johnson-Shoyama Graduate School
of Public Policy
University of Regina
Regina, SK, Canada

ISSN 2365-4961 ISSN 2365-497X (electronic)
Water Governance - Concepts, Methods, and Practice
ISBN 978-3-319-57800-2 ISBN 978-3-319-57801-9 (eBook)
DOI 10.1007/978-3-319-57801-9

Library of Congress Control Number: 2017945273

Printed on acid-free paper

This Springer imprint is published by Springer Nature
The registered company is Springer International Publishing AG
The registered company address is: Gewerbestrasse 11, 6330 Cham, Switzerland

Acknowledgements

Foremost, I would like to express my sincere gratitude to Dr. Joyeeta Gupta for the support, patience, motivation, enthusiasm and great knowledge she has contributed to this book, and my learning. Without her guidance and support, this book would not have been completed. Also my sincere gratitude goes to my promoter, Dr. Hebe Verrest, for her invaluable assistance, input and knowledge that helped strengthen and added depth to this research and book.

I thank all of my colleagues who have worked with me on previous research projects contributing to this book, and a special acknowledgement goes to the invaluable assistance of Jordan Harris (Adapt Chile); Roxana Borquez, PhD candidate, University of London; and Dr. Paula Mussetta (Conicet, Mendoza) who assisted me and contributed greatly to this research.

I wish to present my special thanks to my good friend and colleague, Dr. Harry (Polo) Diaz. Without his support and guidance, this research in Latin America and Canada would not have taken place. His support through these years has been immense.

I would like to show my gratitude to my parents who have encouraged me in pursuing my education. I am sure I would never have made it here without the summers writing books with my father.

I would like to thank Marcel Heemskerk for the preparation of the case study maps and my family – my son Morgan Dynna for creating my Figures, my son Michael Dynna and my husband, Donald Stevens – for their support and understanding while I completed this book.

Contents

Abbreviations

AARD	Alberta Agriculture and Rural Development
ACCD	Agriculture Climate Change Directorate Argentina
ACW	Adaptive Capacity Wheel
ADRMP	Alberta Drought Risk Management Plan
AEMA	Alberta Emergency Management Agency
AESRD	Alberta Environment and Sustainable Resource Development
ASP	Agricultural Social Programme, Mendoza
CBC	Canadian Broadcasting Corporation
CCNS	Climate Change National Strategy, Argentina
CCS	Carbon Capture Storage
CDM	Clean Development Mechanism
CFCs	Chlorofluorocarbons
CNR	National Irrigation Commission, Chile
COP	Conference of the Parties
CSO	Civil Society Organizations
CWB	Canadian Wheat Board
DC	Developing country (Chile and Argentina)
D&f	Drought and flood
DGA	General Directorate of Water, Chile
DGI	*Departamento General de Irrigacion*, Mendoza
DOH	Hydraulic Works Authority, Chile
DRR	Disaster Risk Reduction
FAO	Food and Agriculture Organization
FAPF	Family Agriculture Provincial Forum, Mendoza
FRWIP	Farm Ranch Water Infrastructure Program
G20	Group of 20
GDP	Gross Domestic Product
GEF	Global Environmental Facility
GHG	Greenhouse Gases
GWP	Global Water Partnership
IC	Industrialized Country (Canada)

INDAP	National Institute for Agricultural Development, Chile
INDC	Intended Nationally Determined Contributions
INTA	Institute of Agriculture and Livestock Technology, Mendoza
IPCC	Intergovernmental Panel on Climate Change
IWRM	Integrated Water Resource Management
MOE	Ministry of Environment, Chile
NEC	National Environmental Commission, Chile
NGOs	Non-governmental organizations
NRCan	Natural Resources Canada
ONEMI	National Emergency Office, Chile
PARC	Prairie Adaptation Research Collaborative
PROSAP	Provincial Agricultural Services Program, Argentina
PPWB	Prairie Provinces Water Board
RCMP	Royal Canadian Mounted Police
SBI	Subsidiary Body for Implementation
SCCB	Swift Current Creek Basin
SISS	Superintendent and Sanitary Services, Chile
Sk EmPAct	Saskatchewan Emergency Planning Act
SSRB	South Saskatchewan River Basin
SSRP	South Saskatchewan River Plan
SUCCE	Central Sub-unit Coordination of Emergency, Argentina
UN	United Nations
UNCCD	The United Nations Convention to Combat Desertification
UNDP	United Nations Development Programme
UNECE	United Nations Economic Commission for Europe
UNEP	United Nations Environment Programme
UNFCCC	The United Nations Framework Convention on Climate Change
UNISDR	United Nations Office for Disaster Risk Reduction
USD	United States Dollars
WSA	Saskatchewan Water Security Agency
WSA	Saskatchewan Water Security Act

List of Figures

List of Tables

Chapter 1
Introduction

1.1 Purpose

Agriculture is a fundamental sector providing food, economic output and employment, and a way of life for billions of people. It faces many challenges as a result of climate change, such as changing seasonal patterns, increasing climate variability and more extreme weather events. Extreme weather events and in particular, drought and flood (d&f)[1] are increasing in frequency and intensity (IPCC 2012a, 2014; Parry et al. 2007). The adaptive capacity of agricultural producers to these extreme events needs to be increased in order to reduce their vulnerability to the impacts of drought (through reduced crop and resulting sales revenue) and flood (through damaged crops and equipment). The aim of this research is to increase understanding of policy instruments and goverrnance practices that will increase the adaptive capacity of rural agricultural producers to d&f by strengthening their human, social, economic, technological, and natural capitals.

This book is about adaptive governance. It is about that governance that understands interconnection - interconnection of people (all people), their assets (or capitals) and supportive, effective policy instruments. Most important is the connection of people to their environment. To be adaptive, governance must include the big picture – the big picture not only geographically, but also institutionally (including all formal and informal instititutions). This big picture must include multiple levels of government and mutiple time frames - both short and long term, but also transitionary between different social states, allowing for transformation. Most important for transformation is social learning, which means people must be involved. These adaptive governance conclusions emerge from the theoretical framework constructed in this book and the findings of the four case studies (Alberta and

[1] The impacts of climate change include increasing climate variability, changing seasonal patterns, and increasing frequency and intensity of extreme weather events including d&f (IPCC 2014). This book focuses on d&f and the instruments responding to d&f and climate change. Variability has and will always be present and thus is irrelevant to this book.

M.A. Hurlbert, *Adaptive Governance of Disaster*, Water Governance - Concepts, Methods, and Practice, DOI 10.1007/978-3-319-57801-9_1

Saskatchewan, Canada as well as Coquimbo, Chile and Mendoza, Argentina) of institutionally policy frameworks.

This chapter discusses climate variability and change, d&f, and its impacts on the agricultural sector (see Sect. 1.2). It highlights the key elements of the problem (see Sect. 1.3.1) and the requirement for adaptive governance (see Sect. 1.3.2) outlining what we know and don't know about the problem (see Sect. 1.3.3). The next section (see Sect. 1.4) explains how this research fills some of the gaps in research outlining the research questions posed in order to increase understanding of adaptive governance practices (see Sect. 1.4) and Sect. 1.5 identifies the necessary literatures that require review in Chap. 2. Thereafter, the case study areas are described (see Sect. 1.6), the research focus explained (see Sect. 1.7) and the theoretical perspective outlined (see Sect. 1.8). This chapter concludes by outlining the structure of the book (see Sect. 1.9).

1.2 Impacts of Climate Change on Agriculture and Society

The climate is changing. In the future, increased incidence and intensity of high-temperature extremes and heavy precipitation events are anticipated (Royal Society 2014). The length or number of warm spells or heat waves has increased in many areas of the world and some areas are experiencing more intense and longer droughts (IPCC 2012a: 6, 11). Droughts can be economically devastating: in Canada the most recent widespread drought (2001–2002) produced a USD4.71 billion drop in GDP and was responsible for an estimated 41,000 lost jobs (Wheaton et al. 2010).

At the same time, and often in the same places that drought is being experienced, torrential rains and flooding is increasing in frequency (McHale and Leurig 2012; Comou and Rahmstorf 2012). Japan, Pakistan, and Australia registered record rainfalls in 2011. The economic cost of extreme weather events between 1980 and 2004 is estimated at USD1.4 trillion (of which only one quarter was insured) (IPCC 2012b); much of the huge human cost cannot be quantified (Royal Society 2014). The economic loss calculation is probably low given that losses in terms of cultural heritage, ecosystem services and the informal economy are difficult to measure (*ibid.*). Storms and other weather-related disasters in Latin America and the Caribbean have risen to 571 in 2000–2009 from 356 in 1990–1999, 256 in 1980–1989, and 134 in 1970–1979 (Bello 2015). The number of global disasters has risen over the last four decades to 3496 in 2001–2010 from 2386 in 1991–2000, 1534 in 1981–1990, and 743 in 1972–1980 (WMO 2014). Floods constitute 44% of all disasters and droughts account for 6%. However, droughts result in 35% of disaster related deaths while floods account for 9% (*ibid*).[2]

The state of affairs could get worse. Climate models predict that global average temperature will increase by 1.5 to 4 °C in the twenty-first century (IPCC 2014).

[2]Other disasters include storms, extreme temperature, wildfires, and mass movement (WMO 2014).

There has been a 20% increase in deaths from the period of 1991–2000 compared to 2001–2010 as a result of heat waves, cold spells, drought, storms and floods (WMO 2013). The major climate risk is exposure to extreme conditions and variability (Katz and Brown 1992: 289). Recently the American insurance industry increased its projections of Atlantic hurricane activity by 20% (NAPCO LLC 2011: 7). Countries with low Human Development Index scores are more vulnerable (experiencing far higher mortality rates) (UNDP 2010).

Adapting agriculture to a world with a changed climate and growing world population needs to happen faster as potential catastrophic effects on food production are anticipated (FAO 2011) and climate change has the potential to further marginalize the rural poor, exacerbating their vulnerability. As three quarters of the world's poor live in rural areas and are dependent on farming (Deutsche Bank 2010), supporting these rural agricultural producers is fundamental to the achievement of the Sustainable Development Goals adopted by the United Nations General Assembly in 2015 and the eradication of extreme poverty and hunger. For most low-income countries, agriculture is the most important economic sector utilizing the most labour, containing the majority of the poor and hungry (Cleaver 2013). Rural agricultural development programs achieve the best reductions in rural poverty and hunger (*ibid.*).

1.3 The Problem Definition

1.3.1 Nature of the Problem

The problem of climate change, d&f is daunting. There is **uncertainty and risk** surrounding when and where events of d&f will occur, and their intensity and duration (IPCC 2014). In addition, or perhaps because of this, the science surrounding climate change and adaptation is **contested.** Different scientists predict different future climate scenarios, impacting different variables, resulting in different combinations of variables; they debate the interactions between variables (Holmes and Dinicola 2010).

The magnitude and nature of the impact of d&f events is **systemic**, impacting many social structures including GDP, insurance, agricultural profitability, and food security (see Sect. 1.2). As a result, and in combination with the contested science above, the problem of climate change and adaptation is **complex** (Pahl-Wostl et al. 2013). Further, escalating costs and damages resulting from d&f, uncertainty of science, and systemic impacts, exacerbate the uncertainties surrounding optimal adaptation policy (Haque and Burton 2005). Resolving choices in adaptation efforts are difficult as **values** underpinning adaptation choices are contested. For example: should scarce state funds finance irrigation as an adaptation measure, or provision of drinking water to communities?

It is expected that an increasing variability of climate patterns could go well beyond the existing adaptability range of most people and that the impacts will be borne **unequally**: those living in poverty, women, older adults and young children will have the most difficulty coping with change (Rosenzweig and Tubiello 2006; USGCRP 2009; CCSP 2008; IPCC 2014).

Whether d&f are 'disasters' and whether conflicts and issues surrounding natural resources like water are 'crisis' is often dependent on governance (Bakker 2007) or the pattern of dealing with basic social functions (Lauer et al. 2006) occurring in an institutional[3] context. This institutional context determines the existence of plans to prepare, respond and recover from d&f, and whether adaptive measures are taken to reduce the impact of d&f such that they are not social disasters (Wisner et al. 2005). Governance issues such as these could be exacerbated in the face of increasing climatic variability and uncertainty (Richardson et al. 2009: 12). Poor governance can deteriorate the trust of people in government's management capacities (Conference Board of Canada 2005; Pollution Probe 2007).

Hence, the key elements of the problem surrounding the impacts of climate change, d&f include: (a) uncertainty and risk; (b) systemic nature; (c) complexity; and the fact that (i) science is contested; (ii) values are contested; (iii) and the costs and benefits are inequitably distributed; and (d) poor governance exacerbates the problem.

1.3.2 Why Adaptive Governance Is the Best Solution for the Problem

The increasing intensity, frequency, lives affected, and dollar value of climate change impacts (WMO 2014) and the key elements of the problem (see Sect. 1.3.1) is bringing attention to not only the impacts of climate change but also the necessity of planning in anticipation of these events. It calls for a more concerted response from government regulators and private business (McHale and Leurig 2012). Further, the increasing global disparity in the impact of these events on the poor and marginalized should remain a central issue of analysis (Gupta 2014). For these reasons, tackling issues of adaptation to climate change, d&f need to occur while ensuring inclusive development (Gupta et al. 2015).

[3]An institution is a stable and a collective pattern of dealing with a basic social function (Lauer et al. 2006). It is a set of norms and rules, which define roles and procedures for people, and determines what is appropriate, legitimate and proper. These rules have a degree of permanency and are relatively stable (Homer-Dixon 1999). Formal institutions are represented by water management organizations, government ministries, or legislation and tend to have their rules enforced by a state actor; informal institutions, such as the network of rural agricultural producers formed through the act of having coffee together, embody the socially shared rules that impact behaviour that may be self-enforcing (Helmke and Levitsky 2003). An organization is an institution with an enduring structure represented by a group of individuals bound together by some common purpose to achieve certain objectives (North 1989).

Adaptive governance responds to the key elements of the problem and is the theoretical framework of this research. It recognizes and includes the critical importance of institutions including networks, leadership, and structures in responding to natural resource crisis (Dietz et al. 2003). Adaptive governance addresses risk and uncertainty (see Sect. 2.4.1), applies systemically (recognizing networks and structures), (Berkes et al. 1998) and links these with institutions (see Sect. 2.5.3). It allows for the contestation of science and values (but advances their resolution through public participation) (see Sect. 2.4.2) and creates a space for equitable distribution of resources through participation and inclusive governance by combining these elements. Social learning (learning within social groups through interaction) occurs by addressing the complex systemic problems by changing and deepening understanding by improving routines, questioning assumptions, and changing norms, values and assumptions underpinning strategies and action (see Sect. 2.4.3). Adaptive governance literature seeks to reduce the extreme risk of disasters through disaster risk reduction (see Sect. 2.5.1), and minimize the risk to livelihoods through the livelihoods approach (see Sect. 2.5.2). Lastly, adaptive governance aims to address issues of inequity by addressing its institutional dimension (see Sect. 2.5.3).

Adaptive governance allows for inclusive development, the underlying assumption of this research. Inclusive development has three major components: It includes the natural resource base and ecosystem services on which society depends (here in particular the issues of climate variability and change, water, d&f). Second, it includes issues of access to resources of marginalized people and the allocation of resources, rights, responsibilities and risks within society. Third, inclusive development requires relational approaches to counter the concentration of power and this is addressed here primarily through participatory processes (Gupta et al. 2015).

1.3.3 What We Know and Don't Know

Great strides have been made in the last two decades in understanding challenges confronting humans including climate change, biodiversity loss, and degradation of natural water systems, poverty, energy and food insecurity, and growing social inequality (ISSC/UNESCO 2013; Rockstrom et al. Rockström et al. 2009). Climate change vulnerability assessments have provided information on sensitivity and exposure to changes in climate including the adaptive capacity of systems and populations (Wood 2013). The social and human dimensions of global environmental change has been analyzed by social scientists (see Sects. 1.5, 2.3 and 3.3) and there is much literature on vulnerability and adaptation that informs this research's analysis of the impact on agricultural producers and livelihood assets (see Sect. 2.5.2). Adaptation frameworks have been developed for monitoring and evaluating adaptation, measuring adaptation interventions, setting baselines and targets for monitoring and evaluating, addressing accountability, and learning (Dinshaw et al. 2014).

We know that the institutional governance system is an important component of the adaptive capacity (IPCC 2001: 893–897) of agricultural producers (Hurlbert and

Diaz 2013), as are the resources or assets which agricultural producers have access to in order to build their livelihoods (Moser and Satterthwaite 2008). Nations with "well developed institutional systems are considered to have greater adaptive capacity" and developed countries have a better "institutional capacity to help deal with risks associated with future climate change" (IPCC 2001: 896, 897). Institutions are one of the determinants of adaptation (others are such things as knowledge, technology, education or health) that influence the occurrence and nature of adaptation and thereby circumscribe the vulnerability of systems and their residual impacts (IPCC 2001: 893). Institutions contribute to the management of a community's assets, the community members' inter-relationships, and then in turn their relationships with natural resources. Institutional arrangements shape the processes of endowment and entitlement of rural agricultural producers to livelihood assets (Moser 2009).

The Earth Systems Governance (ESG) Framework (see Sect. 1.3.3) outlined a comprehensive research agenda building on formative prior work of the International Human Dimensions Programme's (IHDP) institutional analysis framework and tackling many questions of governance architecture, adaptiveness, scale, power, norms, and accountability. This framework has informed much research surrounding these concepts and the sectors of water, agriculture, and energy. The framework has contributed to a substantive body of literature and provides 'best practices' (Swanson and Bhadwal 2009; Lim and Spanger-Siegfried 2005) for creating adaptive policies as well as institutional dimensions of adaptive capacity (see Sect. 2.5.3), and a robust body of literature on decision-making in the face of uncertainty (which provides methods, tools and processes for policy makers) (Jones et al. 2014). Some of these tools and processes include integrated watershed management[4] and co-management (Plummer 2009) as well as various models and practices of water governance in relation to topical issues including d&f, sea level rise, extreme climate events, and disaster analysis (Bakker 2007; Bruch et al. 2005).

We know that policy failures repeatedly result from improper framing of issues (Hoppe 2011) and mistaken ideas about behaviour, including that people accept experts' risk analysis, firms support and implement regulations, that free or quasi market incentives work in practice (as they do in theory), and that people act on relevant information (Stern 1993: 1897). To make the issue of climate change even harder to address, we know that adaptation (adjustments in ecological, social, or economic systems in response to actual or expected climatic impacts (Smit and Pilifosova 2001)) is not enough; what is needed is transformation, "a process of change that involves the alteration of fundamental attributes of a system" (ISSC and UNESCO ISSC/UNESCO 2013: 101: Gupta 2014) or learning (Baird et al. 2014). This is the creation of a fundamentally new social system.

[4] Integrated watershed management is a particular strategy of water governance that seeks to model water governance on a geographical watershed basis integrating social, economic, ecological and policy concerns with science by engaging with stakeholders (GWP 2000). Often this management model is touted as ideal and a panacea for all water governance challenges, but is hard to find in practice (Blomquist et al. 2004).

However, there is much we do not know about climate change adaptation. We don't know how decisions are made in the face of uncertainty in the context of governance systems that generate transformations (Hackman and St. Clair 2012). We don't know the common elements needed for transformations, the type of reconfiguration, and who the important actors are in this process (ISSC/UNESCO 2013). We don't know how to confront and reconcile value differences that are often the real origin of conflict between 'objective' experts and government and people (Rayner and Malone 1998: 73).

There is still considerable uncertainty surrounding: (1) the types of governance instruments that achieve resilience or are conducive to adaptation (Eakin et al. 2011; Engle et al. 2011; Huntjens et al. 2012; Eakin and Lemos 2006) and (2) how these governance instruments are framed in policy (Hisschemöller 2005; Adger et al. 2009; Wolf 2012). There is no consensus on what successful adaptation looks like (NRC 2007) or how to define and measure learning in relation to adaptation to climate change (Baird et al. 2014). In fact, the impact of climate change, d&f (IPCC 2012b; Parry et al. 2007) on human well-being and vulnerability[5] is less understood than the transformations of the natural climate system (NRC 2007). There are many ways to approach the problem of increasing risk of d&f and impact on agricultural producers, but some combinations of these have not yet occurred. Scholars have not yet adequately:

(a) Linked an institutional analysis with transformations and social learning (see Sect. 2.4.3) and the existence of double loop (changing assumptions) and triple loop (changing values and norms) learning.
(b) Combined institutional governance assessments with an assessment of impact on livelihood capitals (see Sect. 2.5.2).
(c) Linked adaptive governance with risk, disaster risk reduction, policy framing, social learning and participation (see Sect. 2.4.2).

The methodology of this research explores what we don't know, attempting to fill some of the gaps of knowledge; it also adopts some of these new ways of approaching the problem of climate change (see Sect. 3.3).

1.4 Filling the Gaps

The primary exploratory research question investigates adaptive governance instruments in relation to d&f. The research seeks to determine: **how a theoretical and policy framework (norms, principles, and instruments (including regulatory, economic, suasive, and managerial)) can be designed to build capacity for rural**

[5] Vulnerability is insecurity in the wellbeing of individuals, households and communities, including sensitivity to change, and can be understood as a lack of resilience to changes (environmental, economic, social, and political) that threaten welfare (Moser and Satterthwaite 2008; Moser 1998).

agricultural producers to respond to the increasing likelihood of d&f, defined by uncertainty?

The theoretical framework of this book, grounding this research, is the literature on adaptive governance. Adaptive governance is useful in respect of the area of global environmental change, and in particular in relation to the uncertainty of climate events (especially d&f) as it aims to increase resilience and reduce vulnerability. This book submits that adaptive management, adaptive co-management, anticipatory governance, and disaster risk reduction (see Sect. 2.3) are parts of the adaptive governance system needed for extreme climate events. This study integrates into adaptive governance the concepts of institutional adaptation to climate change (see Sects. 2.2 and 3.3), the livelihoods perspective (see Sects. 2.5.2 and 3.3.7), and risk (see Sect. 2.4.1). In order to address my research question, I use the institutional analysis model of Young et al. (2005) and the Institutional Dimensions of Global Environmental Change (see Sect. 3.2) as foundations to my methodological framework.

Based on this, background information was gathered and a number of subquestions answered. First, the main driving forces affecting rural agricultural producers' livelihoods in the communities relevant to building their capacity to respond to d&f was identified. Second, the main organizations and instruments (formal institutions) which emerged from the legal/policy framework capable of building capacity and responding to d&f at the global level, and at the national and regional level of each study area was identified. Identifying these entailed an iterative study of what existed that was framed in a manner to respond to climate change, d&f, and what agricultural producers accessed in times of d&f, even if it was not specifically framed as relating to climate change, d&f. Gathering this information also allowed for an assessment if any of the environmental governance approaches of adaptive management, adaptive co-management or anticipatory governance were used.

A thorough review of literature on climate change adaptation in the case study area assisted in achieving these first two tasks and also allowed the foundational information on identifying the effectiveness of the main instruments at the community level atachieving their mandated goal in relation to the driving forces. Semi-structured qualitative interviews exploredthis question further, as well as three other questions:

- What effect instruments had on the livelihood capital (human, social, economic, technological, and natural) of agricultural producers in respect of d&f in the study regions?
- Did these instruments advance social learning?
- How can policy instruments and institutions be redesigned to better build capacity for rural agricultural producers to respond to the increasing likelihood of d&f?

The four case studies allowed for an identification of effective instruments and instruments that build the livelihood capital of producers, as well as how these instruments operate witin differing contexts (institutional governance structures, political and socio economic drivers, and social capital etc.).

1.5 The Literature Review

The literature review identified disparate literatures that have applicability in researching climate change, d&f and agricultural policy as it relates to the livelihoods of agricultural producers. These are adaptive governance (Dietz et al. 2003), adaptive management (Bruch 2009; Foxon et al. 2009), adaptive co-management, disaster management, and risk. A literature review was conducted whereby twenty journals over the past 20 years were reviewed for articles on these themes. These journals related to environmental policy, development and politics. In addition, keyword searches of journal databases were conducted using the words adaptive governance, disaster, learning, climate adaptation as well as the case study areas. The journal articles were then sorted, reviewed and findings were integrated.

Chapter 2 reviews the ten literatures. Adaptive governance entails an approach to resource management across multiple scales (Olsson et al. 2006; Berkes et al. 1998) and unites the literatures creating the theoretical framework of this research. I submit that adaptive governance includes adaptive management, adaptive co-management, and anticipatory governance (see Sect. 2.3). Adaptive management emerged in the 1970s as a conceptualization of policy processes that allow for iterative policy monitoring and review for responsive resource management (Williams 2011; Bruch 2009; Plummer 2009) and hence the literature on participation was also reviewed. Adaptive co-management contains many features of adaptive governance but stakeholder involvement in decision making is required (Plummer 2009). Policy framing is an important component of analysis within these literatures serving as a starting point for their unification. Disaster management (disaster risk reduction) focuses specifically on minimizing vulnerabilities and disaster risks through prevention, mitigation and preparedness and relates specifically to d&f (UNISDR 2004; McBean and Rodgers 2010). This necessitated a separate review of the literature surrounding risk. The livelihoods approach and institutional dimension of adaptive capacity operationalize the analysis of the adaptive capacity of agricultural producers. All of these literatures are underpinned by literature conceptualizing risk and have linkages with literature on participation and social learning.

1.6 The Case Studies

This research is a multi-site comparative case study (Bishop 2010). The case study methodology allows for the unpacking and analysis of relationships among mechanisms, contexts, and outcomes driven by those mechanisms identified in the research study (Pawson and Tilley 1997). This unpacking is beneficial for the multi-level institutional study of agricultural producers and d&f. The benefits of the comparative case study are the flexibility of multiple experiments (Yin 2003) that allow iterative model building and comparison. Patterns in the data allow for the addition and extension of the theory and enrich and refine the theoretical framework.

The four cases were selected based on several factors: two sites are in an industrialized country (Canada) and two are in developing countries (Chile[6] and Argentina); previous research connections with the author; similar exposures to d&f linked to climate change; significant irrigated agriculture; and markedly different governance structures. The research sites are river basins in western Canada (Oldman River, Alberta, and Swift Current Creek, Saskatchewan), Chile (Elqui River Basin), and Argentina (Mendoza River). The Canadian, Chilean, and Argentinian river basins represent three large, regional, dryland water basins with significant irrigated agricultural production and strong similarities (see Table 1.1). Climatically sensitive sectors and communities and sensitivity to climate extremes, especially drought in Canada, Argentina, and Chile, characterize these study areas. Alberta and Saskatchewan are considered the most vulnerable regions in Canada to the impacts of projected climate change on water resources (IPCC 2014).

Each case study is illustrated in a chapter and includes a detailed description of each case, and analysing some of the key concepts of this book including the components of adaptive governance present (adaptive management, adaptive co-management, and anticipatory governance), the social learning uncovered, the effectiveness of instruments, and their impact on livelilhood capitals. Multiple sources of data and multiple data collection methods are used (Eisenhardt 1989; Yin 2009). This combination of multi-methods supports triangulation of data to increase the validity of the conclusions (Yin 2003, 2009).

This research capitalizes on four interdisciplinary multi-year research projects that I have participated in that were conducted in the study regions. The first one was a SSHRC (Social Science Research Council of Canada) funded collaborative project between Canadian and Chilean researchers focused on institutional adaptations to climate change (IACC) (see http://www.parc.ca/mcri/).[7] The second project was a project focusing on drought in Saskatchewan and vulnerability of agricultural producers (see http://www.parc.ca/rac).[8] The third project surrounded Water Governance and Climate Change – The Engagement of Civil Society, again funded by SSHRC wherein I studied local watershed advisory groups in the three Prairie Provinces in relation to climate change, the reduction of vulnerability and improving bottom up integrated water governance (See www.parc.ca/vacea/index.php/water-governance).[9] The fourth research project was funded by SSHRC, NSERC,

[6]Chile is a member of the Organization of Economic Cooperation and Development since 2010, but is characterized as a developing country under the Climate Convention of 1992.

[7]Within this project 268 interviews were conducted in Alberta and Saskatchewan, and 86 interviews in the Elqui River Basin of Chile assessing the vulnerability and adaptive capacity of local agricultural producers by a team of social scientists. Thereafter, I conducted a further 51 interviews in Saskatchewan, 49 in Alberta exploring water governance and vulnerability to climate change. A colleague conducted 30 interviews in Chile assessing governance

[8]243 interviews were conducted with agricultural producers in Saskatchewan and I conducted12 interviews with key policy stakeholders with an extensive background in drought and agriculture in Saskatchewan and Canada

[9]One hundred interviews were conducted with local watershed advisory groups.

Table 1.1 Case study descriptions

<table>
<tr><th>River basin, Country</th><th>Location</th><th>Size (km²)</th><th>Agricultural production</th><th>Extreme events</th><th>Irrigated agriculture</th><th>Population</th><th>GDP (Province)</th></tr>
<tr><td>Mendoza, Argentina</td><td>Eastern Andes</td><td>17,821</td><td>Irrigated agriculture, fruits, horticulture, cattle, goats</td><td>Drought 1968, 2010</td><td rowspan="2">Other than goat husbandry approximately 3% of surface is irrigated and agriculture base (Montaña et al. 2005; Santibanez 2015)</td><td>1.28 million (Diaz and Bertranou 2004)</td><td>USD 18,800 nominal (Government of Argentina 2010)</td></tr>
<tr><td>Elqui, Chile</td><td>Western Andes</td><td>9600</td><td>Irrigated agriculture, fruits, cattle, goats</td><td>Drought 2003, 2007</td><td>386,000 (Sauchyn and Santibanez 2010)</td><td>USD 11,351 (purchasing power parity) (Central Bank of Chile 2012)</td></tr>
<tr><td>Oldman, Canada</td><td>Southern Alberta</td><td>26,700</td><td>Irrigated agriculture, grains, pulses, forage, vegetables, cattle</td><td>Drought 2001–2002, Flood 2011</td><td>43,234 of 205,730 Canadian farms are in Alberta and 3817 are irrigated (Statistics Canada 2011, 2005)</td><td>167,383 (Statistics Canada 2008)</td><td>USD 338,166 Millions (Statistics Canada 2014)</td></tr>
<tr><td>Swift Current, Canada</td><td>Southern Saskatchewan</td><td>5592 (Sauchyn and Santibanez 2010)</td><td>Minimal Irrigated agriculture, grains, pulses, forage, vegetables, cattle (Sauchyn and Santibanez 2010)</td><td>Drought1998, Flood 2011</td><td>36,952 of 205,730 Canadian farms are in Sk, and 923 are irrigated (Statistics Canada 2011, 2005)</td><td>1,652,410 (Statistics Canada 2008)</td><td>USD 83,222 Million (Statistics Canada 2014)</td></tr>
</table>

and IDRC, and studied extreme events of drought and flood in Canada, Chile, Argentina, Columbia and Brazil (see http://www.parc.ca/vacea/).[10]

In addition to these projects, social science researchers in Chile and Argentina working in the area of vulnerability and adaptation to climate change assisted with gathering materials, identifying, and interviewing key policy stakeholders. In Chile, Jordan Harris (Adapt Chile) and Roxana Borquez, PhD candidate, University of London provided assistance; in Argentina, Dr. Paula Mussetta (Conicet) contributed greatly to this research.

1.7 Focus

This research is a multi-level institutional case study that focuses on the **agricultural sector**. Agriculture is an important sector as it contributes directly to food security, creates jobs, earns export income, generates funds for savings and investment and also makes indirect contribution to health and prosperity, cultural identity, provides farm tourism and household food security (FAO 2004). Extreme weather events such as drought can reduce agricultural outputs, reducing the Gross Domestic Product of a nation, increase unemployment in many sectors, worsen the balance of trade (Horridge et al. 2005), and sometimes necessitate international aid (Benson and Clay 1998).

This study focuses on the **policy instruments** and their impact on the **livelihoods of agricultural producers**, with a particular focus on the impact of climate change, d&f. A livelihood is defined as, "the assets (natural, physical, human, financial and social capital),[11] the activities, and the access to these (mediated by institutions and social relations) that together determine the living gained by the individual or household" (Ellis 2000: 10). The stock of these assets are significant factors in recovery from environmental shocks and stresses and are influenced by the response and reconstruction process after a disaster or d&f (Moser 2009).

This **multi-level study** involves communities, and local, provincial/regional, and national governments, and international institutions (see Chap. 4). This is necessary as rural agricultural producers function within, and depend upon, local communities (Lonechild and Williams 2008). Jaffe (2003: 4–5) writes:

> Day to day farm work takes place in the context of community – social relations provide much of the invisible web that allows production to take place. Rural (urban) communities

---

[10] In this study in Canada 100 interviews were conducted with local agricultural producers exploring vulnerability to drought and flood, 70 with local governance people exploring the contribution of local governance to the vulnerability of agricultural producers to drought and flood. A similar study led by Paula Musetta was conducted in Argentina.

[11] More specifically defined as: physical capital (equipment, infrastructure, productive resources), financial capital (savings, supplies, credit), human capital (investments in education, health), social capital (an intangible asset including rules, norms, obligations, reciprocity and trust embedded in social relations, social structures and societies' institutional arrangements), and natural capital (stock of environmentally provided assets such as forests, soil, minerals etc.) (Moser 2009).

> in turn derive much of their character from the farm work done there. Simply put rural communities live or die according to the health of the farms that are attached to them, and farms cannot survive without healthy communities.

The local community is often the provider of infrastructure, services, institutions and regulations, which may or may not reduce risks from extreme events. These may take the form of bridges, drainage culverts, roads, hospitals, etc. However, multiple levels of government are also relevant within the institutional governance context including international, national, provincial or municipal governments, non-governmental organizations and private companies.

1.8 A Critical Realist Approach

The research is underpinned (in ontological theory) by the adoption of a broad conception of individual actors as fallible learners, who interact in frequent, repeated and simple situations, often with incomplete information, and are influenced by perceived benefits, costs and reciprocity of relationship (Ostrom 2010). Agricultural producers make day-to-day decisions which impact on their adaptive capacity and exposure to d&f.

This book is informed by critical realism, an appropriate approach for case study social research (Elger 2010: 253). Critical realism posits that a structural reality exists and that it is possible to conceptualize it and make theories in order to describe it (Bhaskar 2010; Danermark et al. 2006). The existence of objective natural and social realities is acknowledged, together with recognition of the socially constructed and fallible character of scientific knowledge (Elger 2010: 253). In this manner, agricultural producers construct their 'risk' or exposure to extreme events of d&f, both objective, natural realities. However, these individual actions and decisions of agricultural producers determine (1) the exposure and adaptive capacity of agricultural producers; and (2) whether natural dry or wet conditions constitute events of d&f. Critical realism differs from the assumption of actors as utility maximizers of Young et al. (2005) (see Sect. 3.2).

Critical realism opposes the traditional dichotomies of positivist and constructivist epistemologies and the associated polarization of quantitative and qualitative methods. The epistemological aim of critical realism is to explain the relationship between experiences, events, and mechanisms, recognizing that there is a reality independent of our knowledge of it; all knowledge is fallible; yet all knowledge is not equally fallible (Bhaskar 2010; Danermark et al. 2006). In assessing policy responses to d&f, and climate change, the significance of the risk of these events, or people and policy maker's construction of the probability of the events, is a reality independent of the natural science and findings of the Intergovernmental Panel on Climate Change (IPCC) or others in this regard.

Critical realism seeks to explain social phenomena (or events and experiences) in terms of the causal powers of particular social mechanisms and their complex interaction in specific contexts. Thus structure and agency are recognized as existing in duality. Structures of social relations are not merely epiphenomena of social interaction, but are characterized by distinct, emergent and enduring properties that constrain or enable different lines of action. The properties may be modified by ensuing individual or collective action (Archer et al. 1998). Individual agricultural producers make decisions within the structure of the suite of policy instruments available to them in the context of their perceptions of natural conditions and the risk of d&f. In this way, the policy instruments relating to climate change and d&f are part of the structure constraining and enabling certain actions. This research explores how agricultural producers engage with policy instruments and how this engagement impacts on their livelihood capitals; this research doesn't assume, for instance, that producers make rational cost benefit decisions in purchasing drought insurance, thereby increasing their economic capital.

1.9 Structure of This Book

This chapter introduced the problem – determining how a theoretical and policy framework can be designed to build capacity of rural agricultural producers and ascertaining the best policy instruments responding to climate change and increasingly frequent extreme events of d&f and the impact on agricultural producers. Questions arise on how to best tackle the problem of climate change and increasing incidents of d&f; these questions are answered in this book by adaptive governance, an adaptive governance that unites multiple literatures and that is comprised of specific policy instruments. Chapter 2 provides a literature review of the theoretical framework of this research - adaptive governance including institutional dimensions of adaptive governance, disaster risk reduction, risk, policy problem framing, participation, and learning. It contributes to the literature by providing a comprehensive and expanded model of adaptive governance. The theoretical framework and literature in Chap. 2 grounds the research methodology and methods in Chap. 3.

This is a multi-level analysis of policy. The international level is reviewed in Chap. 4. The case studies are explored in the subsequent four chapters with Saskatchewan, Canada, in Chap. 5, Alberta, Canada in Chap. 6, Coquimbo, Chile in Chap. 7, and finally Mendoza, Argentina in Chap. 8. The case studies are compared and analysed in Chap. 9, giving insights into how a policy framework can be designed to build capacity for rural agricultural producers to respond to the increasing likelihood of d&f. Chapter 10 contributes to science by reviewing the methodology developed herein and drawing conclusions on a theoretical and policy framework (norms, principles, and instruments (regulatory, economic, suasive, managerial)) designed to build capacity of rural agricultural producers to respond to the increasing likelihood of d&f defined by uncertainty.

References

Adger WN, Dessai S, Goulden M, Hulme M, Lorenzoni I, Nelson DR, Naess LO, Wolf J, Wreford A (2009) Are there social limits to adaptation to climate change? Clim Chang 93(3):335–354

Archer M, Bhaskar R, Collier A, Lawson T, Norrie A (1998) Critical realism, essential reading. Routledge, London/New York

Baird J, Plummer R, Haug CC, Huitema D (2014) Learning effects of interactive decision-making processes for climate change adaptation. Glob Environ Chang 27:51–63. doi:10.1016/j.gloenvcha.2014.04.019

Bakker K (2007) Eau Canada, the future of Canada's water. UBC Press, Vancouver

Bello OD (2015) Planning and disaster risk reduction in Latin America. regional conference on strategies and tools for integrating disaster risk reduction into development planning and financing. United Nations Economic and Social Commission for Asia and the Pacific. Bankgkok, Thailand. Presentation February 18, 2015

Benson C, Clay E (1998) The impact of drought on Sub-Saharan African economies, a preliminary examination. World Bank Technical Paper No. 401. World Bank, Washington

Berkes F, Folke C (eds) (1998) Linking sociological and ecological systems: management practices and social mechanisms for building resilience. Cambridge University Press, New York

Bhaskar R (2010) Contexts of interdisciplinarity: Interdisciplinarity and climate change. In: Bhaskar R, Frnak C, Hoyer K, Naess P, Parker J (eds) *Interdisciplinarity and climate change: Transforming knowledge and practice for our global future*. Routledge, New York

Bishop P (2010). Multi-site case study. In: Mills AJ, Durepos G, Wiebe E (eds) Encyclopedia of case study research. Sage, Los Angeles, pp 587–589

Blomquist W, Schlager E, Heikkila T (2004) Common waters, diverging streams, linking institutions and water management in Arizona, California, and Colorado. Resources for the Future, Washington, DC

Bruch, C. (2009). Adaptive water management: strengthening laws and institutions to Cope with uncertainty. In: Biswas AK et al (eds) Water management in 2020 and beyond, water resources development and management. Springer, Berlin

Bruch C, Jansky L, Nakayama M, Salewicz KA (2005) Public participation in the governance of international freshwater resource. United Nation Press, Tokyo

CCSP (2008) Analysis of the effects of global change on human health and welfare and human systems. A Report by the U.S. climate change science programmes and the subcommittee on global change research. In: Gable JL (ed), Elbi KL, Sussman FG, Wilbanks TJ, (Authors). U.S. Environmental Protection Agency, Washington, DC

Central Bank of Chile (2012) Chile's 2012 Regional GDP in 2008 prices (chain weighted). Available at: http://si3.bcentral.cl/Siete/secure/cuadros/home.aspx. Accessed 03 Apr 2015

Cleaver K (2013) The importance of scaling up for agricultural and rural development, and a success story from Peru. IFAD Occasional Paper 4. Available at: www.ifad.org/pub/op/4.pdf. Accessed 04 Apr 2015

Comou D, Rahmstorf S (2012) A decade of weather extremes. Nat Clim Chang 2:491–496

Conference Board of Canada (2005) Water pressure is building in Canada. *InsideEdge*, Winter 2005, 16

Danermark B, Ekstrom M, Jakobsen L, Karlsson J (2006) Explaining society: critical realism in the social sciences. Routledge, New York

Deutsche Bank Research (2010, February 12) Agribusiness and hunger – threat to global food security drives collaborative business models

Diaz AE, Bertranou A (2004). Systemic study of water management regimes. Global Water Partnership, South America. Available at: http://www.eclac.cl/samtac/noticias/documentosdetrabajo/8/23418/InAr00404.pdf. Accessed 05 Apr 2015

Dietz R, Ostrom E, Stern PC (2003) The struggle to govern the commons. Science 302(5652):1907–1912

Dinshaw A, Fisher S, McGray H, Neha R, Schaar J (2014) Monitoring and evaluation of climate change adaptation: methodological approaches. OECD Enviornment Working Papaers, No. 74, OECD Publishing. http://dx.doi.org/10.1787/5jxrclr0ntik-en. Accessed 29 Mar 2015
Eakin H, Lemos MC (2006) Adaptation and the state: Latin America and the challenge of capacity-building under globalization. Glob Environ Chang 16:7–18
Eakin H, Eriksen S, Eikeland P-O, Oyen C (2011) Public sector reform and governance for adaptation. Implic New Pub Manag Adapt Capa Mexico Norway 47:338–351
Eisenhardt K (1989) Building theories from case study research. Acad Manag Rev 14(4):532–550
Elger T (2010) Critical Realism. In: Mills AJ, Durepos G, Wiebe E (eds) Encyclopedia of case study research, vol 1. Sage, Los Angeles
Ellis F (2000) Rural livelihoods and diversity in developing countries. Oxford University Press, Oxford
Engle NL, Johns OR, Lemos MC, Nelson DR (2011) Integrated and adaptive management of water resources: tensions, legacies and the next best thing. Ecol Soc 16(1):19–30
FAO (Food and Agriculture Organization of the United Nations) (2011) Potentially catastrophic climate impacts on food production over the long-term March 31, 2011. Food and Agriculture Organization of the United Nations website. Available at: http://www.fao.org/news/story/en/item/54337/icode/. Acesssed 15 July 2015
FOA (2004) Socio-economic analysis and policy implications of the roles of agriculture in developing countries. Roles of Agriculture Project. Research Programme Summary Report 2004. FOA, Rome
Foxon R, Reed MS, Stringer LC (2009) Governing long-term social-ecological change: what can the adaptive management and transition management approaches learn from each other? Environ Policy Govern 19:3–20
Government of Argentina (2010) Indec Census. Available at: http://www.censo2010.indec.gov.ar/resultadosdefinitivos_totalpais.asp. Accessed 03 Apr 2014.
Gupta J (2014) The history of global climate governance. Cambridge University Press, Cambridge
Gupta J, Bersgsma E, Termeer CJAM, Biesbroek GR, van den Brink M, Jong P, Klostermann JEM, Meijerink S, Nooteboom S (2015) The adaptive capacity of institutions in the spatial planning, water, agriculture, and nature sectors in the Netherlands. Mitig Adapt Strateg Glob Chang. doi:10.1007/s11027-014-9630-z
GWP (Global Water Partnership) (2000) Integrated water resources management. TAC Background Papers, No. 4. Stockholm, Sweden
Hackmann H, St. Clair AL (2012) Transformative cornerstones of social science research for global change. International Social Science Council, Paris
Haque E, Burton I (2005) Adaptation options strategies for hazards and vulnerability mitigation: an international perspective. Mitig Adapt Strateg Glob Chang 10:335–353
Helmke G, Levitsky S (2003) Informal institutions and comparative politics: a research agenda. Working Paper #307, Kellogg Institute
Hisschemoller M (2005) Participation as knowledge production and the limits of democracy. In: Maassen S, Weingart P (eds) Democratization of expertise? Exploring novel forms of scientific advice in political decision-making, pp 189–208
Holmes RR Jr, Dinicola K (2010). 100-Year flood–it's all about chance. U.S. Geological Survey General Information Product 106:1
Homer-Dixon T (1999) Environment, scarcity, and violence. Princeton University Press, Princeton
Hoppe R (2011) The governance of problems, puzzling, powering and participation. The Policy Press, Bristol
Horridge M, Madden J, Wittwer G (2005) The impact of the 2002-2003 drought on Australia. J Policy Model 27(3):285–308
Huntjens P et al (2012) Institutional design propositions for the governance of adaptation to climate change in the water sector. Glob Environ Chang 22:67–81
Hurlbert M, Diaz H (2013) Water governance in Chile and Canada – a comparison of adaptive characteristics. Spec Feat Govern Adapt Ecol Soc 18(4):61–76

IPCC (2001) Climate change 2001: impacts, adaptation, and vulnerability technical summary, a report of working group II of the intergovernmental panel on climate change 2001, intergovernmental panel on climate change. WMO and UNEP, Geneva

IPCC (2012a) Summary for Policymakers. In: Field CB, Barros V, Stocker TF, Qin D, Dokken DJ, Ebi KL, Mastrandrea MD, Mach KJ, Plattner G-K, Allen SK, Tignor M, Midgley PM (eds) Managing the risks of extreme events and disasters to advance climate change adaptation, A special report of working groups I and II of the intergovernmental panel on climate change. Cambridge University Press, Cambridge/New York, pp 1–19

IPCC (2012b) Managing the risks of extreme events and disasters to advance climate change adaptation. In: Field CB, Barros V, Stocker TF, Qin D, Dokken DJ, Ebi KL, Mastrandrea MD, Mach KJ, Plattner G-K, Allen SK, Tignor M, Midgley PM (eds) A special report of working groups I and II of the intergovernmental panel on climate change. Cambridge University Press, Cambridge/New York

IPCC (2014) Summary for policymakers. In: Edenhofer O, Pichs-Madruga R, Sokona Y, Farahani E, Kadner S, Seyboth K, Adler A, Baum I, Brunner S, Eickemeier P, Kriemann B, Savolainen J, Schlömer S, von Stechow C, Zwickel T, Minx JC (eds) Climate change 2014, mitigation of climate change. Contribution of working group III to the fifth assessment report of the intergovernmental panel on climate change. Cambridge University Press, Cambridge/New York

ISSC/UNESCO (2013) World social science report 2013: changing global environments. OECD Publishing and UNESCO Publishing, Paris

Jaffe J (2003) Introduction. In: Diaz J, Stirling (eds) Farm Communities at the Crossroads: Challenges and Resistance. CPRC Press, Regina

Jones RN, Patwardhan A, Cohen SJ, Dessai S, Lammel A, Lempert RJ, Mirza MMQ, von Storch H (2014) Foundations for decision making. In: Field CB, Barros VR, Dokken DJ, Mach KJ, Mastrandrea MD, Bilir TE, Chatterjee M, Ebi KL, Estrada YO, Genova RC, Girma B, Kissel ES, Levy AN, MacCracken S, Mastrandrea PR, White LL (eds) Climate change 2014: impacts, adaptation, and vulnerability. Part A: global and sectoral aspects, Contribution of working group II to the fifth assessment report of the intergovernmental panel on climate change. Cambridge University Press, Cambridge/New York, pp 195–228

Katz RW, Brown BG (1992) Extreme events in a changing climate: variability is more important than averages. Climate Change 21(3):289–302

Lauer RH, Lauer C, Abramson Z, Auger J (2006) Social problems and the quality of life. McGraw Hill Ryerson, Toronto

Lim B, Spanger-Siegfried E (eds) (2005) Adaptation policy frameworks for climate change: developing strategies, policies and measures. UNDP-GEF. Cambridge University Press, Cambridge

Lonechild G, Williams R (2008) ACRE Final subcommittee report. Redefining community. In: Community action committee on the rural economy. Enterprise Saskatchewan, Regina

McBean GA, Rodgers C (2010) Climate hazards and disasters: the needs for capacity building. Wiley interdisciplinary reviews. Climate Change 1-6:871–884

McHale C, Leurig S (2012) Stormy future for U.S, Property/casualty insurers: the growing costs and risks of extreme weather events. A Ceres Report, September 2012. MA, Boston

Montaña E, Torres LM, Abraham EM, Torres E, Pastor G (2005) Los espacios invisibles. Subordinacion marginalidad y exculsion de los territories no irrigados en las tierras secas de Mendoza, Argentina. *Región y Sociedad* 17(32)

Moser CON (1998) The asset vulnerability framework: reassessing urban poverty reduction strategies. World Dev 26(1):1–19

Moser C (2009) A conceptual and operational framework for pro-poor asset adaptation to urban climate change. Paper presented to the 5th urban research symposium, June 28–30, 2009. Available at: http://siteresources.worldbank.org/INTURBANDEVELOPMENT/Resources/336387-1256566800920/moser.pdf. Accessed 14 July 2015

Moser C, Satterthwaite D (2008) Towards pro-poor adaptation to climate change in the urban centres of low- and middle-income countries. Human settlements discussion paper series theme: climate change and cities – 3. International Institute of Environment and Development

(IIED). Paper presented for the World Bank's Social Development Department Workshop on the "Social Dimensions of Climate Change" held on March 5, 6 2008 at the World Bank, Washington, DC

NAPCO LLC (2011) The impact of changes to the RMS U.S. Hurricane Catastrophe Model. Available at: http://www.napcollc.com/articles/JuneReviewRMSHurricane.pdf. Accessed 14 Apr 2015

North D (1989) Institutions and economic growth: a historical introduction. World Dev 17:1319–1332

NRC (National Research Council) (2007) Evaluating progress of the U.S. climate change science program: methods and preliminary results, committee on strategic advice on the U.S. Climate Change Science Programmes (Washington, DC), 3–4

Olsson P, Gunderson LH, Carpenter SR, Ryan P, Lebel L, Folke C, Holling CS (2006) Shooting the rapids: navigating transitions to adaptive governance of socio-ecological systems. Ecol Soc 11(1):1–18

Ostrom E (2010) Beyond markets and states: polycentric governance of complex economic systems. Am Econ Rev 100(3):641–672

Pahl-Wostl C, Giupponi C, Richards K, Binder C, de Sherbinin A, Sprinz D, Toonen T, van Bers C (2013) Transition towards a new global change science: requirements for methodologies, methods, data and knowledge. Environ Sci Pol 28:36–47

Parry ML, Canziani OF, Palutikof JP, van der Linden PJ, Hanson CE (eds) (2007) Summary for policymakers. In: Climate change 2007: impacts, adaptation and vulnerability. Contribution of working group II to the fourth assessment report of the intergovernmental panel on climate change. Cambridge University Press, Cambridge, pp 7–22

Pawson R, Tilley N (1997) Realistic evaluation. Sage, London

Plummer R (2009) The adaptive co-management process: an initial synthesis of representative models and influential variables. Ecol Soc 14(2):24–40

Probe P (2007) Towards a vision and strategy for water management in Canada, final report of the water policy in Canada: national workshop series. Polis Project on Ecological Governance, Victoria

Raynor S, Malone EI (1998) Social science insights into climate change. In: Rayner S, Malone EI (eds) Human Choice & Climate Change, Volume 4, What have we Learned? Battelle Press, Columbus

Richardson K et al (2009) Synthesis report, climate change global risks challenges and decisions. University of Copenhagen, Copenhagen

Rockström J, Steffen W, Noone K, Persson Å, Chapin FS III, Lambin E, Lenton TM, Scheffer M, Folke C, Schellnhuber H, Nykvist B, De Wit CA, Hughes T, van der Leeuw S, Rodhe H, Sörlin S, Snyder PK, Costanza R, Svedin U, Falkenmark M, Karlberg L, Corell RW, Fabry VJ, Hansen J, Walker B, Liverman D, Richardson K, Crutzen P, Foley J (2009) Planetary boundaries: exploring the safe operating space for humanity. Ecol Soc 14(2):32. Available at: http://www.ecologyandsociety.org/vol14/iss2/art32/. Accessed 5 June 2015

Rosenzweig C, Tubiello FN (2006) Developing climate change impacts and adaptation metrics for agriculture. Paper produced for the global forum on sustainable development on the economic benefits of climate change policies, July 6–7 Paris, France. Available at: www.oecd.org/environment/cc/37117548.pdf. Accessed 4 Mar 2015

Royal Society (2014) Resillience to extreme weather. The Royal Society Science Policy Centre report 02/14 issued November 2014. The Royal Society, London. Available at: https://royalsociety.org/~/media/policy/projects/resilience-climate-change/resilience-full-report.pdf. Accessed 30 Mar 2015

Santibanez F (2015) Study area: Limar, Coquimbo Region, Chile. .Available at: www.noveltis.com/desurvey/interface/site_region.php. Accessed 13 Sept 2013

Sauchyn D, Santibanez F (2010) Vulnerability to climate extremes in the Americas. Application for funding to the Tri-Council (SSHRC, NSERC, IDRC). Available online at: www.parc.ca

Smit B, Pilifosova O (2001) Adaptation to climate change in the context of sustainable development and equity. In: McCarthy JJ, Canzianni OF, Leary NA, Dokken DJ, White KS (eds) Climate change 2001: impacts, adaptation, and vulnerability – contribution of working group II to the third assessment report of the intergovernmental panel on climate change. Cambridge University Press, Cambridge UK

Statistics Canada (2005) Table A proportion of irrigated farm operations, 2005. Queen's Printer, Ottawa

Statistics Canada (2008) 2006 Community Profiles. Available at: http://www12.statcan.ca/english/census06/data/profiles/community/Index.cfm?Lang=E. Accessed 03 Apr 2015

Statistics Canada (2011) Census of agriculture. Queen's Printer, Ottawa

Statistics Canada (2014) Gross domestic product, expenditure-based, by province and territory. Available at: http://www.statcan.gc.ca/tables-tableaux/sum-som/l01/cst01/econ15-eng.htm. Accessed 03 Apr 2015

Stern PC (1993) A second environmental science: human-environment interactions. Science 260(June 25):1897–1899

Swanson D, Bhadwal S (eds) (2009) Creating adaptive policies, a guide for policy-making in an uncertain world. Sage, New Dehli

UNDP (2010) Reducing disaster risk: a challenge for development. UNDP, New York

UNISDR (2004) Living with risk: a global review of disaster reduction initiatives. United Nations International Strategy for Disaster Reduction Secretariat. Available at: http://www.unisdr.org/we/inform/publications/657. Accessed 17 July 2015

USGCRP (2009) Global climate change impacts in the United States. In: Karl TR, Melillo JM, Peterson TC (eds) United States global change research program. Cambridge University Press, New York

Wheaton E, Kulshreshta S, Wittrock V (2010) Assessment of the 2001 and 2002 Drought Impacts in the Prairie Provinces, Canada. In: Sauchyn D, Diaz H, Kulshreshtha S (eds) Chapter 16: The new normal. Canadian Plains Research Center, Regina

Williams BK (2011) Adaptive management of natural resources, framework and issues. J Environ Manag 92:1346–1353

Wisner B, Blaikie P, Cannon T, Davis I (2005) At risk: natural hazards, people's vulnerability and disasters. Routledge, New York

WMO (2014) Atlas of mortality and economic losses from weather, climate and water extremes (1970–2012). WMO-No. 1123. Geneva, Switzerland

WMO (World Meteorological Organization)(2013) The global climate of 2001–2010: a decade of climate extremes. WMO-No-1103, Geneva

Wolf J (2012) Climate change adaptation as a social process. In: Ford JD, Berrang-Ford L, Li H, Gupta J, Van Dijk MP (eds) Climate change adaptation in developed nations China's drought strategies in rural areas along the Lancang river. Water Policy

Wood L (2013) Compendium of lessons learned from ARCC climate change variaiblity assessments. African and Latin American resilience to climate change project. U.S. Agency for International Development, Bulinton

Yin R (2003) Application of case study research. Sage, Thousand Oaks

Yin RK (2009) Case study research: design and methods, 4th edn. Sage, Thousand Oaks

Young OR, Agrawal A, King LA, Sand PH, Underdal A, Wasson M, Revised edition prepared by Schroeder H (2005) Science plan. Institutional dimensions of global environmental change. International Human Dimensions Programme on Global Environmental Change (IHDP), Bonn

Chapter 2
Adaptive Governance (Management, Co-management and Anticipatory)

2.1 Introduction

This chapter explores and unites literature surrounding adaptive governance. First, the particularities of governance, in particular adaptive governance, are evaluated (see Sect. 2.2). Building on why adaptive governance best addresses - the problem of d&f in relation to agricultural producers (see Sects. 1.3.2 and 2.2), this chapter examines: How has adaptive governance evolved (see Sect. 2.3.2)? How do the adaptive governance, adaptive management, adaptive co-management and anticipatory governance literatures relate (see Sect. 2.3.3)? What are the key gaps in the literature of adaptive governance in relation to climate change and extreme events (see Sect. 2.3.4) and how can these gaps be filled (see Sect. 2.4)? This chapter takes an ideal-typical approach to defining the above terms; in fact authors of each term (adaptive management, adaptive co-management and anticipatory governance) often stretch these to take new elements into account, thus creating confusion about the breadth of each of these terms.

Risk and uncertainty is a key element of the problem (see Sect. 1.3.1). Since adaptive approaches focus on uncertain events, they all deal with the notion of risk. Hence I explore the construction of risk (see Sect. 2.4.1), which is also an important aspect of policy problem framing. Since adaptive governance starts with problem framing, I link these theories together (risk, problem framing and adaptive governance) (see Sect. 2.4.2). Moreover, a critical feature of adaptive governance is stakeholder participation and learning. These elements have also been integrated into this theoretical framework of adaptive governance.

An important question surrounding governance of d&f is what is the purpose of adaptation? Governance is for whom and to achieve what end? Here the literature relating to agricultural producer livelihoods, and reduction of vulnerability to d&f is integrated into adaptive governance to clarify what end state the governance system is to achieve (see Sect. 2.5.2). Adaptive governance also seeks to make the governance system more adaptive and resilient in order to better strengthen all of its

M.A. Hurlbert, *Adaptive Governance of Disaster*, Water Governance - Concepts, Methods, and Practice, DOI 10.1007/978-3-319-57801-9_2

interconnecting relationships and patterns of decision-making. To do this, the literature on best institutional practices of adaptive governance was incorporated into this research (Gupta et al. 2010).

The unification of this literature allows for the development of two new models: the model of adaptive governance and problem structuring (see Sect. 2.4.2 and Hurlbert and Gupta 2016), and adaptive governance and split ladder of participation model (see Sect. 2.4.3 and Hurlbert and Gupta 2015). These models are used to enrich the adaptive governance literature and research methodology (see Sects. 2.5 and 3.2). The implications of adaptive governance literature for this research are outlined in Sect. 2.6.

2.2 Why Adaptive Governance?

Addressing extreme events like d&f calls for adaptive governance (see Sect. 1.3.2). Adaptive governance goes beyond existing resource management approaches by incorporating ecosystem dynamics, resilience[1] and complexity theory (Berkes and Folke 1998), addressing links across different scales (Folke et al. 2005, 2007; Gunderson and Holling 2002; Olsson et al. 2006), at multiple levels of governance (Lebel et al. 2006). In addition, adaptive governance recognizes that although people are mostly rational social actors, their knowledge is imperfect, unevenly distributed, and they apply different evaluation criteria to different institutional settings depending on communications or transparency (Hatfield-Dodds et al. 2007).

Adaptive governance allows for the systemic nature of the problem of d&f, which reflexive governance, focusing on politics and technological transitions in transitions management and path dependency, does not (Vob and Bornemann 2011). Polycentric governance is, "complex, modular systems where differently sized governance units with different purpose, organizations, spatial location interact to form together a largely self-organized governance regime" (Pahl-Wostl 2009: 357). Although polycentric governance allows for multiple points of decision-making and influence (Huitema et al. 2009), its focus loses track of the individual agricultural producer and multiple levels of structure surrounding this key unit of analysis that is the focus of this research. As governments increasingly respond to frequent extreme climate events (IPCC 2014), a broader focus than just government, instruments, and policy making is required (IPCC 2001; Hurlbert et al. 2009) (the focus of many models of environmental management).[2]

[1] Resilience is the persistence or robustness of a system in the face of disturbance or shocks and understanding how communities can innovate in the face of complex, fast or slow changes, drawing on institutional memory and their ability to self-organize, recombine structures and processes, renew systems and find new ways (Folke 2006).

[2] Management is conceptually different than governance. Management refers to "the processes of decision making, coordination and resource deployment that occur within a given institutional setting assuming no change in rules and norms" (Hatfield-Dodds et al. 2007: 3).

Adaptive governance aims, inter alia, to resolve the problem of d&f (see Sect. 1.3.1). It aims to enhance the adaptive capacity of people to cope with and reduce the risks to their livelihoods (see Sect. 2.5.2) and their vulnerability to disaster (d&f). It also aims to enhance the adaptive capacity of institutions (see Sect. 2.5.3) thereby enhancing the resilience of agricultural producers and their communities.

2.3 Defining Adaptive Governance

2.3.1 Introduction

This section assesses what constitutes adaptive governance and its components, and how do they inter-relate? Governance entails the interactions among formal and informal institutions, i.e. traditions, norms, rules, processes and structures that determine how people in societies make decisions and share power, exercise responsibility and ensure accountability (Lebel et al. 2006: Raik and Decker 2007; Fabricius and Cundill 2014). Governance refers to political, legal, social, economic and administrative institutions that develop, manage, and distribute societal goods (such as water) (Rogers and Hall 2003) involving public, private and civil society organizations that practice and implement the norms, programmes, regulations, and laws relevant to this exercise (de Löe et al. 2009; Armitage et al. 2009).

2.3.2 Evolution of Adaptive Governance

In the 1970s practices of 'adaptive management' emerged with the seminal publication on adaptive management of Holling (1978)[3] to assist government and policy makers with new conceptualizations and processes for establishing, formulating, monitoring and evaluating policy in response to ever increasing environmental and public pressures (Bruch 2009; Williams 2011). Walters (1986) provided a more complete technical treatment and Lee (1993) expanded the conceptual framework to include social and political dimensions. Table 2.1 presents this evolution and the following sections discuss the literature.

To this framework, concepts of participation and social learning have been added to arrive at adaptive co-management (Plummer and Armitage 2007; Folke et al. 2003; Armitage et al. 2007; Olsson et al. 2004). Co-management initially focused on dualistic power sharing between the government and local users (Berkes 1994). Subsequently it emphasised a broader spectrum of actors and the process of adaptive co-management expanded to include continuous problem solving (Plummer 2006). The co-management concept was then merged with adaptive management as

[3] Although earlier articulations occur in Beverton and Holt (1957) who described adaptive decision making in fisheries, without referring to it as adaptive management.

Table 2.1 Evolution of adaptive governance literature

<table>
<tr><td colspan="3">Adaptive Governance – includes the theories below and emerged in the late 1990s. It merged resilience and complexity theory, and reflected the trend of moving from government to governance. It increasingly includes elements from the other three types of governance (Berkes and Folke 1998; Gunderson and Holling 2002; Folke et al. 2005; Olsson et al. 2006)</td></tr>
<tr><td>Theory</td><td>Development</td><td>References</td></tr>
<tr><td rowspan="3">Adaptive management</td><td>Began in 1970s</td><td rowspan="3">Holling (1986), Bruch (2009), and Williams (2011)</td></tr>
<tr><td>Evolved from theories of resilience, as well as business, experimental science, systems theory, and industrial ecology.</td></tr>
<tr><td>Has expanded in recent decades to add social and institutional conditions facilitating the transfer of science in the decision-making process and the idea of social learning</td></tr>
<tr><td rowspan="3">Adaptive co-management, adaptive collaborative management</td><td>Began in late 1990s</td><td rowspan="3">Plummer and Armitage (2007), Folke et al. (2003), Armitage et al. (2007) and Olsson et al. (2004)</td></tr>
<tr><td>Merged adaptive management (the learning dimension) with collaborative management (the linkage dimension).</td></tr>
<tr><td>Emerged at Center for International Forestry Research in 1997 and also independently</td></tr>
<tr><td rowspan="2">Anticiptory governance</td><td>Emerged in 2001</td><td rowspan="2">Quay (2010) and Cole (2001)</td></tr>
<tr><td>Emerged from scenario planning and adaptive management literature; deliberative democracy and complexity theory</td></tr>
</table>

an innovation in natural resource governance under conditions of change, uncertainty and complexity (Armitage et al. 2007: 2).

Anticipatory governance followed, focusing on scenario planning (Quay 2010; Chi 2008; Cole 2001) and adaptive governance (Berkes and Folke 1998; Gunderson and Holling 2002; Folke et al. 2005; Olsson et al. 2006); both include aspects of participation and social learning. Adaptive governance emerged at the local level as a loosely coordinated array of pragmatic responses to manifest failures of scientific management (Brunner and Lynch 2010: 5). It combines the literature surrounding resilience and complexity theory (Berkes and Folke 1998) and management of the commons (NRC 2001). This chapter argues that adaptive governance includes the elements of adaptive management, adaptive co-management, and anticipatory governance. However adaptive governance broadens the lens beyond these approaches to include the study of multiple social structures, interconnected resources and actors (Brunner and Steelman 2005; Berkes and Folke 1998).

Table 2.2 Definition of adaptive governance, adaptive management, adaptive co-management, and anticipatory governance

Theory	Definition and approach	References
Adaptive management	A regulatory environment that utilizes the adaptive management methodology to address uncertainty. First the problem and desired goals are identified; next the system boundaries and context are ascertained. Hypotheses are developed and tested which leads to the implementation of policy strategies and monitoring of results. This all occurs in a management cycle of continuous monitoring, assessment and revision. Process is emphasized over goals	Foxon et al. (2009: 7), Swanson and Bhadwal (2009) and Bruch (2009: 91) and Pahl-Wostl (2007a)
Adaptive co-management[a]	Flexible community-based systems of resource management tailored to specific places and situations supported by various organizations at different scales and levels. A long-term management structure permits stakeholders to share management responsibility for natural resources and to learn from their actions. Combines learning dimensions of adaptive management and linkage dimensions of co-management	Plummer (2009), Olsson et al. (2004: 75) and Armitage et al. (2007)
Anticipatory governance	A flexible decision framework that uses a wide range of possible futures to prepare for change and guides current decisions towards maximizing future opportunities and minimizing future threats. Quay finds the roots of anticipatory governance in scenario planning and adaptive management and Farooque in deliberative democracy and complexity theory	Quay (2010: 506), Chi (2008), (Cole 2001) and Farooque (n.d.)
Adaptive governance	A range of political, social, economic, and administrative systems that develop, manage and distribute a resource in a manner that promotes resilience through collaborative, flexible, and learning –based issue management across different scales.	Berkes and Folke (1998), Gunderson and Holling (2002), Folke et al. (2005) and Olsson et al. (2006)

[a]Synonymous with adaptive collaborative management (Plummer and Armitage 2007) and community based natural resource management with a focus on adaptive capacity (Armitage 2005)

2.3.3 Comparing Components of Adaptive Governance

I argue that adaptive governance is composed of adaptive management, adaptive co-management and anticipatory governance. However, before expounding on this, this section assesses how each is defined in the literature and how they inter-relate. Table 2.2 defines the inter-related and overlapping approaches of each. Table 2.3 outlines the focus and scope of adaptive management, adaptive co-management, anticipatory governance, and adaptive governance.

Adaptive management is a scientific or technical approach involving an iterative process of resource management aimed at reducing uncertainty over time by using

Table 2.3 Characteristics of co-management, adaptive management, adaptive co-management, anticipatory governance, and adaptive governance

Characteristic	Adaptive management	Adaptive co-management	Anticipatory governance	Adaptive governance
Focus on establishing linkages	Learning by doing in a scientific and deliberate way	Establishing horizontal and vertical linkages to carry out joint learning by doing	Linking resource planning with scenario and strategic planning (Chi 2008; Margerum 2005)	Linkages between institutions and social networks as well as human and environmental interactions (Berkes and Folke 1998)
Implementation steps	Detailed process of transition envisioned requiring building trust, strengthening mechanisms, etc. (Bruch 2009)	Stages involved in bringing actors together, building shared values, developing pathways, and undertaking tasks (Plummer 2009), preparing for change, seizing window of opportunity (Olsson et al. 2004)	Deliberative democracy branch envisions specific planning combined with empowerment, leadership, and integration (Farooque n.d.)	Prescriptive implementation plans not given. A new generation (Scholz and Stiftel 2005) or vision of governance institutions generally cited all practicing the institutional prescriptions of adaptive governance (see e.g. Hatfield-Dodds et al. 2007)
Temporal scope	Medium – to long-term: multiple cycles of learning and adaptation)		Medium to long -term with changes implemented in modules and implemented as needed as the future unfolds (Mean et al. 2010)	Is challenged to expand into broader spatial and temporal scales building knowledge, incentives, and learning capabilities into institutions and organizations of governance that then allow adaptive management at the local, regional, and global scales (Berkes and Folke 1998)
Spatial scope	Focus on managers' needs and relationships	Multi-scale, across all levels, with attention to needs and relationships of all partners	Focus on managers' needs and relationships (Fishkin 2009; Farooque n.d.)	
Focus on capacity building	Focus on resource managers and decision makers	Focus on all actors: 'two (or more) to tango'	Focus on resource managers and decision makers utilizing scenario planning (Margerum 2005)	Improving adaptive capacity of governance including people, local governance bodies, government etc. (Hattfield-Dodds et al. 2007)

management actions as experiments to test policy and subsequent system monitoring (Walters 1986). Adaptive management is a structured process that seeks to 'learn by doing' (Doremus 2001) or to learn to manage by 'managing to learn' (Bormann et al. 1999), and adapting based on what is learned (Walters and Holling 1990).

The literature is diverse in what exactly is adaptive management, the centrality of learning and experimentation, the role of uncertainty, and differences in who adapts to what (Dewulf et al. 2007). It is consistent in its appropriateness for a regulatory resource environment where there is uncertainty about environmental dynamics, impacts of proposed activities, and future demands (Bruch 2009). The literature is also generalizable in respect of the process of adaptive management (Pahl-Wost 2007a) that includes (a) an initial assessment phase of system boundaries and context, problems and desired goals (McLain and Lee 1996; Folke et al. 2003; Williams 2011); (b) which occurs in collaboration with stakeholders (Williams 2011); (c) adoption of a provisional legal, policy, and institutional framework; (d) ongoing monitoring and collection of information; (e) periodic assessment to determine effectiveness, and (f) then appropriate modification of the framework (Foxon et al. 2009; Bruch 2009).

There has been an expanding dynamic focus of adaptive management over time with more recent applications including stakeholder involvement or participation (Norton 2005), and social learning (Pahl-Wostl 2007a).[4] Adaptive management remains tied to the operational level (Pahl-Wostl 2007a) focused primarily on the first element of management, the identification of strategy (Hatfield-Dodds et al. 2007: 4).

The concept of adaptive co-management is intricately woven, highly nuanced, and often difficult to dissect (Plummer and Armitage 2007). No single model of adaptive co-management emerges in the literature (Plummer 2009). Adaptive co-management is defined as (a) flexible community-based systems of resource management (b) tailored to specific places and situations, (c) supported by and working with various organizations at different levels (Folke et al. 2003). (d) It is a long-term management structure that (e) permits stakeholders to share management responsibility within a specific system of natural resources, and (f) to learn from their actions through a dynamic, self-organizing process of learning by doing (Folke 2006). Adaptive co-management has both horizontal and vertical connections managing complex adaptive systems (Huitema et al. 2009; Olsson et al. 2004). Adaptive co-management systems build on open institutions and learning (Shannon and Antypas 1997), shared rights, management power and responsibility, and may involve multiple institutional links with user groups, communities, government agencies and non-governmental organizations (Armitage et al. 2007; Olsson et al. 2006).

[4] However, with the increased emphasis on stakeholder involvement within the iterative review process, adaptive management becomes conceptualized as a polycentric style of governance that doesn't have a single center (Foxon et al. 2009). Thus adaptive management becomes synonymous with adaptive governance. Bertha (2014) clearly distinguishes adaptive management from collaborative adaptive management and the inclusion of stakeholders.

Uncertainty exists whether adaptive co-management is a forged institutional arrangement or an emergent one (Berkes 2007: 320). Plummer (2009) and Olson et al. (2004) envision a transformation from top-down government approaches to adaptive co-management involving a change to participatory community engagement which is either led by a local policy entrepreneur (Olsson et al. 2004) or determined by exogenous variables (government mandates, social and political context including culture, knowledge systems, power) and endogenous variables (leadership, human capital, and social capital) (Plummer 2009).

Anticipatory governance is a new concept[5] and has two streams. The first employs a management language using a flexible decision making framework with a wide range of possible futures to prepare for change and to guide current decisions towards maximizing future alternatives or minimizing future threats (Quay 2010: 496; Margerum 2005) through embracement of scenario planning (Chi 2008; Cole 2001; Quay 2010). It is described as a "system of institutions, rules and norms that provide a way to use foresight for the purpose of reducing risk, and to increase capacity to respond to events at early rather than later stages of their development" (Feurth 2009: 29). Flexible actions are recommended that can be broken into modules and implemented as needed as the future unfolds (Hallegatte 2009). 'No regrets' strategies and 'worst case' strategies can be developed; many strategies may address a number of possible future scenarios (Mean et al. 2010). Indicators of change should be monitored on a regular basis and decisions to implement anticipated adaptation strategies considered in light of actual trends (Quay 2010: 499).

The second stream of anticipatory governance employs a deliberative democracy and complexity theory model focusing on building capacity to respond to unpredictable tasks (Guston 2008) by incorporating scenario development (indicators and performance measures for anticipatory knowledge, FHSC 2009). This anticipatory knowledge must also be used and applied through collaborative consensus building in a deliberative democratic process (Dryzek 2000) about the future based on a collective and shared visioning process building on the principles of deliberative democracy (Fishkin 2009).

The first stream focuses more on management while the second stream on governance, albeit much of the literature appears to use these terms interchangeably.[6] Management refers to the processes of decision-making, coordination and resource deployment that occurs within a given institutional setting, assuming no change in rules and norms (Hatfield-Dodds et al. 2007: 3). Governance (see Sect. 1.3.1) is broader as it deals with multiple people, groups, organizations, across multiple lev-

[5] Several scholars have used the term 'anticipatory' in relation to policies, without defining it (Lebel et al. 2010; Hertin and Berkhout 2002).

[6] There is overlap between the adaptive management and adaptive governance literatures. First, the literature slips from utilizing the term 'adaptive management' into 'adaptive governance' without acknowledging the difference (Lee 1993; McLain and Lee 1996). For instance Booher and Innis (2010: 35) seemingly use the terms interchangeably and in the end these authors arrive at a form of collaborative governance. Pahl-Wostl uses a framework of transitioning to adaptive management as a means of achieving adaptive governance (2010: 512). Several authors use the term adaptive management but employ a methodology of adaptive governance (Eakin et al. 2011).

els and scales. This distinction also assists in the consideration of **adaptive governance**, as adaptive governance refers to a socio-ecological system where policy is implemented in modest steps in both a top down (although fragmented) and bottom up manner (Brunner and Steelman 2005).

Similar to adaptive management, adaptive governance operates when change is occurring with respect to a resource (Berkes and Folke 1998), knowledge is incomplete, and science is uncertain (Brunner and Steelman 2005) but also in a situation which includes multiple ways of knowing and understanding issues (*ibid.*). Adaptive governance has particular applicability to intricately, closely interconnected human and biophysical systems (Dietz et al. 2003) extending resource management with an understanding of ecosystem dynamics (Berkes and Folke 1998).

Although the adaptive governance literature is not always consistent and in agreement, key characteristics of adaptive governance which emerge are collaborative, flexible, and learning-based issue management across different scales (Folke et al. 2005; Gunderson and Holling 2002; Olsson et al. 2006) aiming to build flexible institutions and social networks in multi-level governance systems striving to improve adaptive capacity to deal with uncertainty, surprise and external drivers (Berkes and Folke 1998). It includes continuous learning by interpreting and responding to ecosystem feedback and stakeholder participation in policy making to explore and understand uncertainty (Hatfield-Dodds et al. 2007). Hence, I believe that adaptive governance is the broadest term, including adaptive management, adaptive co-management and anticipatory governance. Figure 2.1 presents the inter-relationships between these literatures respecting disaster, d&f.

Adaptive governance represents the entire institutional governance system, while the other three governance approaches focus only on particular aspects or only on one resource system. Adaptive governance is especially applicable to systemic problems such as climate change, factoring it into thousands of local problems, each of which is more tractable scientifically and politically than the global problem (Brunner and Lynch 2010). Adaptive governance is broader than the other three approaches in spatial and temporal scope (Berkes and Folke 1998). It includes the social and human context for applying adaptive management (Dietz et al. 2003) as it deals with the complex human interactions that have been obstacles to the implementation of adaptive management (Lee 1993; Gunderson 1999) but are increasingly built into the adaptive management literature (Bertha 2014). These obstacles are fundamentally institutional rather than technical, as institutions are built on major premises and long-held beliefs that are deeply embedded in educational systems, laws, policies, and norms of professional behaviour (Stankey et al. 2005). Adaptive governance includes networks, leadership, and institutional structures (Carpenter et al. 2006; Dietz et al. 2003).

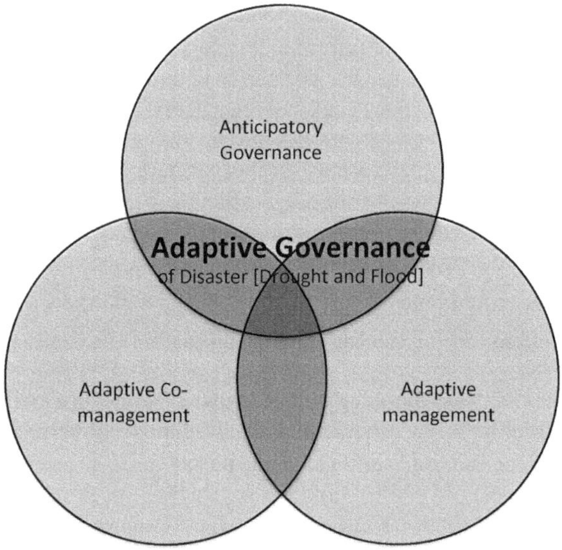

Fig. 2.1 Adaptive governance of disaster: d&f

2.3.4 *Shortcomings of Adaptive Governance*

The adaptive governance literature has weaknesses:

1. Adaptive governance merges resilience and complexity theory (Berkes and Folke 1998), but better theories of probing risk, uncertainty, and complexity are required;
2. Better methods for exploring issues of complexity – incomplete knowledge, contested science, and multiple ways of knowing are required (Brunner and Steelman 2005) beyond polycentric decision making (Huitema et al. 2009) or simply incorporating ecosystem dynamics (Berkes and Folke 1998);
3. Solutions aren't offered for resolution of conflicts in multiple ways of knowing and disputes of science or values in relation to adaptation;
4. Although broadly appropriate for many situations, the broadness of adaptive governance makes it lacking in detail and specifically lacking in a method of application or analysis; and
5. The adaptive governance literature requires a clearer outcome to measure. Measuring whether an institution or policy is adaptive is useful, but often dependent on normative judgments of researchers ranking institutional practices; more is needed.

These weaknesses identified above, are remedied in this book by expanding the adaptive governance literature to include risk (see Sect. 2.4.1), including policy framing to better explore complexity (see Sect. 2.4.2), probing further into participation and learning to resolve conflicts (see Sect. 2.4.3), and enhancing a methodology by providing clearer outcomes to measure (see Sect. 3.3). This enhanced model of adaptive governance better responds to the problems of uncertainty and risk, the systemic nature of the problem, its complexity, and competing values and science.

2.4 Enhancement of Adaptive Governance

2.4.1 Risk

How should risk be understood in the context of adaptive governance, d&f? A situation that involves exposure to danger is a risk (Oxford Dictionary 2010). Risk is defined as the effect of uncertainty (inadequate or incomplete knowledge or understanding resulting in deficiency of information) on objectives (ISO 2009). The likelihood of a certain event occurring (likely to rare) and its consequences (insignificant, minor, moderate, major, or catastrophic) are determined (Standards Australia 2009; Deloach 2000). Arguably in the last three categories of consequences, a risk becomes an emergency, and potentially if actualized, a disaster. From this perspective risk is objective and calculable.

The opposite viewpoint is the constructivist perspective of risk that finds risk to be socially constructed (with factors such as experiences, emotions, attitudes, and knowledge shaping one's perception of risk) (Hubbard 2009; Renn 2011) which also informs the categorizing of an event as serious enough to be regarded as a hazard. Events are 'risks' only insofar as they are recognized as such by a set of socially ascribed decisions together with a component of calculative practice. As a result, risks only enter the 'real' world, when they are cognized and then managed as part of a social process (Zinn 2008).

Combining the realist and constructivist approach to risk (a "weak constructionist" approach) posits that risk is both a real and socially constructed reality and talk about risks mutually influence and produce each other (Zinn 2008: 8). Even though risks are necessarily real, they are socially selected, transformed and debated. This approach is exactly the approach of critical realism, the ontological theory of this research (see Sect. 1.8). The reality of a risk is a prerequisite for persistent debates and activities on risk, but their politicization is culturally determined (*ibid.* 6).

The level of preparation for risk avoidance is based on how policymakers construct or envision the risk, which is predicated on environmental, legal, economic factors, and people's judgements about risk acceptability, their feelings, dread, and outrage factors surrounding it (Slovac et al. 2004). Recognizing this duality of reality and social construct is consistent with critical realism (see Sect. 1.8). The definition and construction of what is or is not a risk is important in planning and responding to climate change (Preston et al. 2009). There is a scientific and objective knowledge surrounding climate change and d&f, but there is a very identifiable cultural construct of climate science (Von Storch 2009; Zinn 2008).

Laypeople don't accept realist, objective risk (utilized in the risk management approach); risk is mediated by geography and the impact of culture (Preston et al. 2009; Zinn 2008; Von Storch 2009). Institutions and policy instruments on climate change and d&f may or may not exist, or vary from country to country.

The adaptive governance literature responds to risk as outlined in Table 2.4.

The approaches of adaptive co-management and adaptive management adopt a realist approach (being objectively for a broader conception of risk). In anticipatory

Table 2.4 Approaches to risk

Literature	Approach to risk	Handling of risk
Adaptive governance	Uncertainty of not only ecosystem, but also human institutions (Scholz and Stifel 2005).	Being prepared for change and surprise; enhancing adaptive capacity to deal with disturbance (Folke et al. 2005).
Adaptive co-management	Recognition of different epistemologies and multiple uncertainties by making decisions as close to user as possible and allowing for pluralism (Armitage et al. 2007).	Problems of resources increasingly negotiated by agreement with stakeholders and less by experts (Brunner and Steelman 2005).
Adaptive management	Technocratic approach to risk and uncertainty orchestrated by decision or policy maker (Brunner and Steelman 2005).	Hypothesis testing of policy response in learning-by-doing, scientific, deliberative way, with focus on managers' needs and relationships (Armitage et al. 2007).
Anticipatory governance	Risk is operationalized by presentation of scenarios and reduction to one plan and one process through prioritization, albeit public participation may be involved (Quay 2010).	Reflexivity developed through scenario planning and adaptive management considering wide range of possible futures; Strategic plans developed and monitored (Quay 2010).

governance and adaptive governance a broader approach to risk recognizes that risk can be based on peoples' diverse perceptions (a subjective construction), as well as the realist approach (an objective construction). These differing perspectives are resolved through their respective processes. Recognizing the differing perspectives allows for an understanding of differing perspectives and values, deepens the debate surrounding contested issues, and ultimately allows for their resolution.

2.4.2 *Model of Adaptive Governance and Problem Structuring*

As adaptive governance entails closely interconnected human and biophysical systems across multiple scales rife with risk and uncertainty, it is constituted by multiple interconnected problems. This section will expand on the analysis of these problems through policy framing and structuring.

Policy framing refers to how a particular problem, or risk, is constructed. A policy 'problem'[7] is a gap between a current situation and a more desirable future one (Hoppe 2011: 23). Implicit in this definition is the social and political construct which articulates that a particular state of affairs is undesirable, that a particular 'risk' to society exists, and that a more desirable future state (in accordance with

[7]A policy is specifically defined as a "course of action or principle adopted or proposed by a government, party business or individual" (OECD 2012: n.p.).

values, norms, and goals) can be attained by governmental (Hisschemöller and Hoppe 1996) or governance action. 'Reality' is thus bound up with the 'perception' (Carroll 1988: 1; see Sect. 2.4.1) of actors, and specifically those actors with the power to determine the policy agenda (Hisschemöller and Gupta 1999). The existence of institutions, policies, and instruments in relation to climate change and d&f is determined by whether these events are identified as risks and are therefore, policy problems.

Many organizations clearly relate to d&f response including disaster response organizations, emergency measures departments, and flood control units. Also pertinent are crop insurance agencies, home and business insurers, government agricultural departments, irrigation associations, and environmental departments. The determination of which organizations and institutions are relevant (and then which instruments and policies) is a function of how 'problems' of extreme events of d&f are framed and structured. D&f may or may not be framed as events of climate change. Although the science is well established that increasing frequency and intensity of d&f can be anticipated with climate change (Sauchyn et al. 2010; Villalba et al. 2009), people may or may not frame these events within the climate change policy problem framework.

Climate change is a policy problem, which can be broken down into several smaller policy problems. Adapting to future climate change includes the policy problems of preparing municipal infrastructure for increasing frequency and intensity of floods as well as improving the adaptive capacity of rural producers to drought. An issue typology, or how a policy problem is structured or framed and the resulting policy's form and content, determines how policy makers and the public construct meaning around the problem and determines how the problem is analysed (Lebel et al. 2010). Table 2.5 structures the problems of climate change, and responding to d&f. It reflects on how the literature of responding to risk (see Sect. 2.4.1) lines up with the various sub problems of climate change, d&f.

Structured problems are problems where there is substantive agreement on norms, principles, ends and goals surrounding a policy problem, as well as agreement on relevant and required knowledge inherent in solving the problem. These are policy problems where a realist, objective risk exists. These problems are largely specialist and determined by technical experts or bureaucratic specialists who are guardians of the general interest. An example of a structured environmental problem is identifying the cost-effectiveness of different crop practices to reduce soil erosion, or determining the costs and benefits of expanding an irrigation project (Batie 2008: 1177). Table 2.5 displays structured problems such as reducing soil erosion and restoration of wetlands to prevent flooding as embedded problems of adapting to climate change. Figure 2.2 illustrates the relationship between structured and unstructured problems in relation to agreement on values/norms and certainty of science.

A moderately structured policy problem occurs when policy makers have either some agreement on norms, principles, ends and goals in defining a future state, or some agreement on the relevant and required knowledge inherent in solving the problem, but not agreement on both norms as well as knowledge (Hoppe 2011:

Table 2.5 Climate change policy problem framing

Responding to climate change			
Adapting to climate change			Mitigating climate change
Embedded problems:			
Responding to drought	Responding to flood	Responding to disaster	Reducing GHG Emissions and thereby reducing the long-term need to continuously deal with the risks of climate change
Increasing water resilience – enhancing water security	Protecting species habitat; Enhancing water security	Preparing and practising emergency plans	
Preventing soil erosion	Constructing dams and weirs; Restoring wetlands; Protecting property and person; Increasing water storage capacity of rural landscape	Arranging emergency housing and financial assistance; Providing aid in the form of water, food, services	
Assessing expanded irrigation; Increasing water storage capacity of rural landscape			

73–75). Certain aspects of adaptation policy can be moderately structured problems:

(a) preparing municipal infrastructure for increasing frequency and intensity of floods (where knowledge is sound, but who pays is tenuous; quadrant Q4 of Fig. 2.2);
(b) improving the adaptive capacity of rural producers to respond to drought (where assisting this group of actors is agreed on as a valuable goal, but the science of future climate scenarios is uncertain (Jones and Preston 2011). See Q2 of Fig. 2.2.

Unstructured problems are those in which uncertainty exists in respect to the norms and values at stake as well as in regard to the certainty of required and available knowledge located in Q3 of Fig. 2.2 (Hoppe 2011: 73). These are akin to 'wicked problems', social messes, or untamed problems, which are dynamically complex, ill-structured, public problems. Mitigating climate change is such a problem. The direct and indirect causes and effects of the problem are extremely difficult to identify and model. Wicked problems tend to be intractable and elusive because they are influenced by many dynamic social and political factors as well as bio-physical complexities (Rittel and Webber 1973). In addition, wicked problems tend to be connected to or are symptoms of other problems (Carroll et al. 2007). These are problems where there is a socially constructed or a weak constructionist perception of risk (see Sect. 2.4.1).

Adaptation to climate change is potentially a wicked, unstructured problem in Q3. When there is both a lack of consensus on social values as to whether adaptation to climate change should occur, as well as disagreement on the means and ends for adapting to climate changes (Adger et al. 2009: 342–342), it is an unstructured

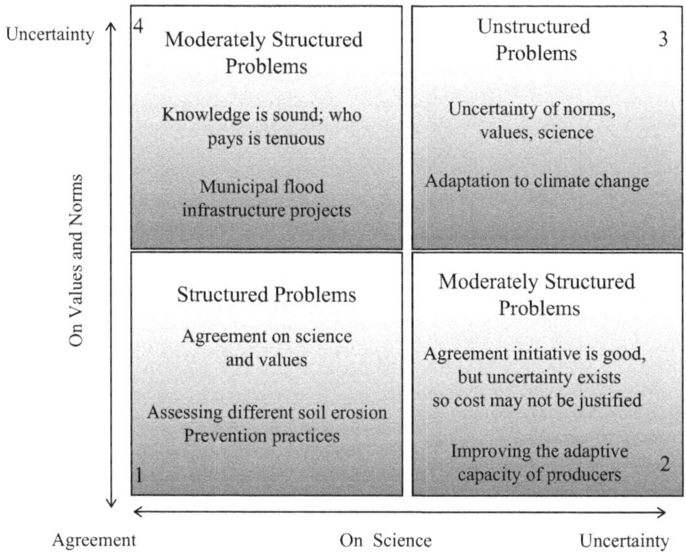

Fig. 2.2 Four types of policy problems identified in policy analysis (Source: Adapted from Hisschemöller and Hoppe 1996: 44; Hisschemöller and Gupta 1999: 157; Hurlbert and Gupta 2016)

problem. However, components of this policy problem may become structured or moderately structured policy problems, as discussed above (Hoppe 2011).

One determinant of how and whether the component policy issues set out in Table 2.5 are reflected in the institutional context (organizations, policies, instruments of a case study) is how the risk of the problem (flood, drought, climate change) is constructed within its institutional setting. If there is no construct that droughts or floods are a localized 'risk', a policy problem will not exist and therefore no policy or instruments will be required. The perception of climate change science (as uncertain, or certain) and the norms and values surrounding adaptation (as agreed to, or not) will determine where on the policy spectrum the various policy problems reside, and be an indication of the policy that may or may not exist in each case study in relation thereto.

Adaptive governance responds to the unstructured, wicked problems embracing the suite of policy instruments in Q3 of Fig. 2.2, but includes the totality of all the other quadrants. Adaptive governance is improved with a full consideration of certainty in Q1, and uncertainty (in science, norms and values Q2, 3, 4) inherent in climate change science, policy making and framing. Adaptive management (see Sect. 2.3) responds to scientific uncertainty and is located on the bottom left hand side of the policy problem figure. The process of adaptive management can tackle moderately structured problems here by interrogating uncertainty in science in Q2; adaptive co-management processes effectively address moderately structured problems addressing differing values and norms in Q4; Anticipatory governance

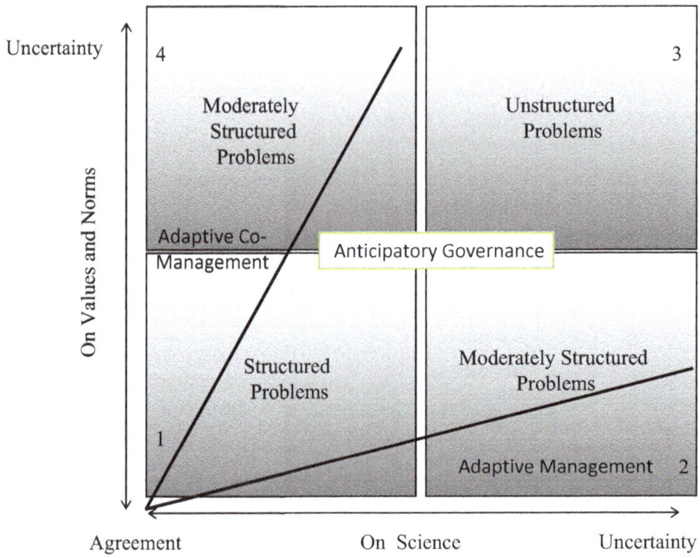

Fig. 2.3 Model of adaptive governance and problem structuring (Source: Hurlbert and Gupta 2016)

processes (see Sect. 1.1) reside just on the inner central dimension of wicked problems tackling targeted unstructured problems entering Q3. See Fig. 2.3 above.

2.4.3 *Adaptive Governance and the Split Ladder of Participation (Inclusive Development)*

Learning is an important theme in adaptive management, adaptive co-management, and adaptive governance.[8] The adaptive management process is premised on 'learning by doing' (Lee and Lawrence 1986) or 'managing to learn' (Bormann et al. 1999) through its strategic intervention allowing hypotheses to be tested through experimentation (Holling 1978; Walters 1986). The inclusion of public participation in both adaptive management and adaptive governance is also based on the specific goal of learning because of the iteration of data collection and stakeholder input (Bruch 2009 103; Pahl Wostl et al. 2007).

Although there is little consensus on the theoretical basis and meaning of social learning (Reed et al. 2006: 58), here it implies learning in and with social groups through interaction (Argyris 1999; Siebenhuner 2008) or collaboration, and organization, and learning which occurs in networks of interdependent stakeholders

[8] The risk management literature does envision post disaster reviews of preparedness and response plans (Henstra 2011), a form of learning.

(Mostert et al. 2007). In this research context, it becomes a multi-step social process in which individuals and organizations need to learn how to manage different framings of issues while raising awareness of climate risks and opportunities, exploring policy options and institutionalizing new rights, responsibilities, feedback and learning processes for climate adaption in the long term (Tabara et al. 2010). Learning is more than just providing more knowledge on climate impacts to policy makers. Learning is an important part of engaging with uncertainty (Newig et al. 2010: 339–340; Bijlsma et al. 2011: 54). It has been found that the most democratic form of participation maximizes both the opportunity for information flow (in both directions from the public to policy makers) and social learning through iterative exchanges of information between all parties (Rowe and Frewer 2000). The potential benefits of social learning include an increased ability to cope with informational uncertainty, assistance with building consensus for criteria to monitor and evaluate environmental systems, reduction of resource conflicts, enhanced fairness of decisions, and increased adaptive capacity (Lebel et al. 2010: 334–335).

Social learning should demonstrate that a change in understanding has taken place in the individuals involved, as well as demonstrate that a change has occurred which goes beyond the individual and becomes situated within wider social units, communities of practice, social networks or institutions (Reed et al. 2006). The concepts of single, double, and triple loop learning help to assess the depth of social learning (Argyris 1999; Fabricius and Cundhill 2014). Single loop learning refers to incremental changes such as the improvement of existing routines or instrumental change in strategy. This change occurs within the constraints of current societal norms and beliefs. Double loop learning involves questioning the assumptions and mental models underpinning strategies and action. Triple loop learning occurs when values and norms that underpin assumptions are questioned and reflected upon leading to a change in underlying world views; deeper understanding of the context, power dynamics and values influencing the management of natural resources supports this learning (Pahl-Wostl 2009, Gupta 2014). In triple-loop learning, learning goes beyond insight and patterns to context and a shift in understanding occurs such that a transformation occurs (Tschakert and Dietrich 2010). I add to this model the concept of zero loop learning (bottom left hand corner of Fig. 2.4) where there is not only an absence of learning, but a regression of learning. Here, previous practices that built resilience are forgotten, or practices that are ultimately harmful to building livelihood capitals are implemented.

Often the literature romanticizes participation and learning, without addressing when participation should inform policy problems and what type of learning is needed in relation to what type of problem. The traditional ladder of participation sees a linear progression of participatory goals from manipulation to self-management. However, Fig. 2.4 below illustrates the split ladder of participation (Hurlbert and Gupta 2015) that splits the ladder at the top and at the bottom. It links the ladder of participation to adaptive governance and the interconnections of policy problem, participation and learning. For structured policy problems in Q1, technocrats/bureaucrats generally take decisions in the public interest, interacting with the public on occasion to educate or be educated, and occasionally using traditional

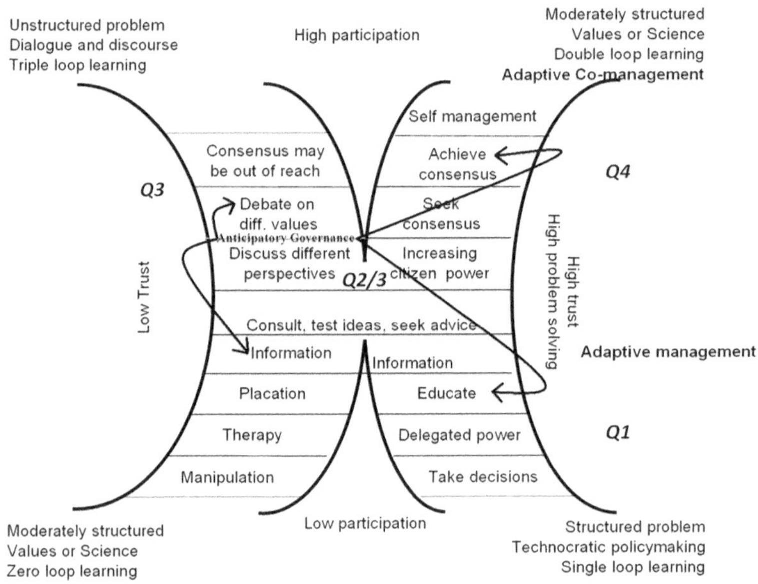

Fig. 2.4 Adaptive governance and the split ladder of participation (Source: Adapted from Hurlbert and Gupta 2015)

adaptive management. Moderately structured problems appear in Q2 and Q4. The bottom left is the area illustrated by low levels of trust and participation of people resulting in outcomes of manipulation, therapy or placation; conversely Q4 illustrates high levels of trust and participation tackling moderately structured problems in which double loop learning may occur. Q3 is where unstructured problems reside and triple loop learning is required. Here there is very low trust, but ironically much more trust is needed for triple loop learning. These issues may through participation be converted into moderately structured problems or structured problems through dialogue and participation (such as extreme drought resulting in local drinking water committees and sharing). In the process more trust may be created.

It is in the top two thirds of the Figure that both adaptive governance and inclusive development occurs, where inclusive development implies addressing social well being issues, living within the Earth's carrying capacity and relational aspects of addressing power politics through greater empowerment through participation (Gupta et al. 2015). The arrows depict the process of anticipatory governance: iterative, participatory problem solving. It is here that institutional practices respond to uncertainty and sometimes contested values and science. These processes include participation in a focused beneficial way, depending on the policy problem. In this manner marginalized voices are heard and taken into account and trust is built such that the United Nation's Development Programme's goals of integrating the standards and principles of human rights (participation, non-discrimination and accountability) are achieved (UNDP 2015).

2.5 Goals of Adaptive Governance

In the literature, the goal of adaptive governance is to enhance the adaptive capacity of institutions (in order to deal with uncertainty, surprise and external drivers (Berkes and Folke 1998)). This research adds two new goals: reducing risk and increasing livelihood capitals. Each goal is discussed in turn.

2.5.1 Adaptive Governance Aims to (a) Reduce the Risk of d&f

Specific literature has developed on managing and responding to disaster. Disaster risk reduction (DRR)[9] is a systematic approach to identifying, assessing, and reducing the risks of disaster (Alexander 2002). DRR is defined as:

> The conceptual framework of elements considered with the possibilities to minimize vulnerabilities and disaster risks throughout a society, to avoid (prevention) or to limit (mitigation and preparedness) the adverse impacts of hazards, within the broad context of sustainable development (UNISDR 2004: 17).

Although there is still some variability in method and theoretical underpinning in relation to DRR (e.g., Preston et al. 2009), Funfgeld and McEvoy (2011) identify three stages in the literature. In the first stage, hazards are viewed as disasters in an objective, deterministic manner based on economic rationalism; in the second stage by emphasizing the physical protection of assets from hazards and engineering solutions to reduce the impact of natural hazards on communities; the third DRR stage emphasizes strengthening people's capacity to absorb and recover from hazards- by reducing the negative effects of development practices on vulnerability. A fourth stage identified in the late 1990s posits that structural social inequality, not nature or technology, creates hazards. This stage calls for reducing vulnerability and mainstreaming DRR in sustainable development (Funfgeld and McEvoy 2011). This fourth stage involves a shift from managing disasters for the reduction of disaster risks (UN/ISDR 2005) to recognizing the complexity and systemic nature of the problem. This evidences **triple loop social learning** (a marked change in values and norms underpinning assumptions). This research engages with all stages: from the technocratic Q1 response to d&f, to and including addressing the Q3 systemic nature of climate change.

[9]Termed hazard research by some (Funfgeld and McEvoy 2011) and often used interchangeably with disaster risk reduction (Moser 2009).

2.5.2 *Adaptive Governance Aims to (b) Increase Livelihood Capitals of Agricultural Producers*

An extension of the vulnerability approach specifically to disasters is the livelihood asset adaptation approach. This approach includes and extends the vulnerability literature focusing on the micro level of an individual livelihood and its capacity or capability to adapt to change or absorb shock and build resilience. Resilience is based on adaptation capabilities of individuals and households and deeply influenced by factors ranging from: damage or destruction to their homes and other assets, constraints on their livelihood prospects, and including the social psychological effects of deprivation and exclusion (Moser 2009). In expounding on livelihood capital analysis, Moser (2009) reviews literatures on climate change adaptation (DRR, climate risk management, climate change adaptation, climate change vulnerability resilience, community-based adaptation, capital-based vulnerability and adaptation) and notes that all approaches focus to a varying extent on capitals; from this, a capital-focused framework is developed (see Sect. 3.3.7) identifying the most important capital assets for individuals, households and communities, which reduce vulnerability to a range of hazards (Moser 1998). Vulnerability is linked to a lack of, or an erosion of physical, financial, human, social and natural capital. These capitals are a significant factor in self-recovery after disaster (*ibid.*), and institutions play an important role in the process whereby the capitals of individuals and households are protected (de Haan and Zoomer 2005). For these reasons, livelihood capitals are included in relation to actors in the multi-level institutional analysis of this research.

2.5.3 *Adaptive Governance Aims to (c) Enhance the Adaptive Capacity of Institutions*

Six dimensions of adaptive governance[10] that governance regimes should have in order to facilitate adaptation and enhance the adaptive capacity of institutions include **Variety**, learning, room for autonomous change, leadership, resources and fair governance (Gupta et al. 2010). In respect of the dimension of variety there are four criterion: room for a variety of problem frames reflecting different opinions and problem defiinitions (Nooteboom 2006); involvement of different actors at different levels, within different sectors for a multi-actor/level/sector dimension (Pahl-Wostl 2009); availability of a wide range or diversity of policy option to address a

[10] Note that the terminology of dimensions of adaptive governance and the content of these principles are by no means consistent. Some authors term them 'evaluative criteria' (Ostrom 2011) or even 'elements' of adaptive institutions (Mollenkamp and Kastens 2009). This book uses the adaptive capacity dimensions of Gupta et al. (2010: 5). The discussion in some cases is generic and applies to institutions in general (Gupta et al. 2010; Gunderson and Holling 2002; Olsson et al. 2006) and in other cases to specific institutional systems such as water governance (Mollenkamp and Kastens 2009; Huntjens et al. 2012).

particularl problem (Verweij and Thompson 2006); and lastly a redundancy or duplication of measures, back-up systems that may not necessarily be cost effective (Wick and Sucliffe 2001).

The **Learning** dimension includes five criteria including trust (mutual respect (Pelling and High 2005)), single loop learning (ability to improve routines based on past experience (Pelling et al. 2008)), double loop learning (changed underlying assumptions of institutional patterns (Ormond 1999)), discussion of doubts (openness to uncertainties, monitoring and evaluation of policy experiences (Pahl-Wostl 2009)), and institutional memory (via monitoring and evaluation of policy experiences over time (Ostrom 2005)).

Room for autonomous change includes three criteria: Continuous access to information (data, institutional memory and early warning systems to individuals (Folke et al. 2005; Milman and Short 2008)); Acting according to plan (especially in relation to disasters (Smit et al. 2000)); and Capacity to improvise (in relation to self-organization and fostering social capital (Armitage 2005)).

Leadership is expressed through three types: visionary which is long term and reformist, entrepreneurial which leads by example (Malnes 1995), and collaborative, or between different actors (Pahl-Wostl 2007b).

Resources refer to those in relation to authority (legitimate forms of power that may or may not be embedded in constitutional laws (Biermann 2007)), human resources of expertise, knowledge and labour (Nelson et al. 2010), and financial resources that make policy measures achievable (Nelson et al. 2010).

Lastly **Fair governance** has four components: Legitimacy or public support, equity in relation to institutional fair rules, responsiveness to society (Biermann 2007), and accountability in relation to procedures (Biermann 2007).

Institutional systems demonstrating strong performance in relation to these dimensions are more resilient, and enhance the adaptive capacity of the system to a greater degree than institutions not demonstrating these dimensions (Gupta et al 2010; Mollenkamp and Kastens 2009; Gunderson and Holling 2002; Olsson et al. 2006).

2.6 Implications of Adaptive Governance for Climate Related d&f Research

This chapter has reviewed the literatures of adaptive governance utilizing the ideal typical approach of each. The evolution, inter-relationships, and respective processes of adaptive governance were reviewed. Adaptive governance is argued to subsume another three literatures: adaptive management, adaptive co-management, and anticipatory governance. The weaknesses of adaptive governance were addressed by creating an expanded definition of adaptive governance by exploring dimensions of risk, participation and learning, and adding ideas surrounding policy framing. Two new models of adaptive governance in relation to risk and policy problem were developed: the adaptive governance and policy structuring model (Fig. 2.3; Hurlbert and Gupta 2016) and the adaptive governance and split ladder of participation model (adapted from Hurlbert and Gupta 2015) (Fig. 2.4).

The expanded models inform the methodology of this research. As adaptive governance is conceived in institutional terms, the methodological framework of multi-level institutional analysis (see Sect. 3.3) is appropriate. The expanded definition of adaptive governance informs this methodology providing criteria for assessment of institutions. The adaptive governance and policy structuring model informs the analysis of institutions and instruments (see Sect. 3.3.2); the adaptive governance and split ladder of participation model informs the analysis of impacts on actors (see Sect. 3.3.6). The goal of adaptive governance of improving livelihoods informs the methodology by providing measurement of impact on society (see Sect. 3.3.7). Lastly, the institutional dimensions of adaptive governance leverage this information and provide analysis for improving the policy framework to increase the capacity of agricultural producers to adapt to d&f (see Sect. 3.3.8). The methodology described in Chap. 3 provides for consideration of the interconnection of local agricultural producers to their communities, their governments, their informal and formal institutions in an inclusive manner all of which is theoretically grounded in adaptive governance.

References

Adger WN, Dessai S, Goulden M, Hulme M, Lorenzoni I, Nelson DR, Naess LO, Wolf J, Wreford A (2009) Are there social limits to adaptation to climate change? Clim Chang 93(3):335–354

Alexander D (2002) Principles of emergency planning and management. Oxford University Press, New York

Argyris C (1999) On organizational learning. Blackwell, Oxford

Armitage D, Berkes F, Doubleday N (2007) Adaptive co-management, collaboration, learning and multi-level governance. UBC Press, Vancouver

Armitage D (2005) Adaptive capacity and community-based natural resource management. Environ Manag 35(6):703–715

Armitage DR, Plummer R, Berkes F, Arthur RI, Charles AT, Davidson-Hunt IJ, Diduck AP, Doubleday NC, Johnson DS, Marschke M, McConney P, Pinkerton EW, Wollenburg EK (2009) Adaptive co-management for social-ecological complexity. Front Ecol Environ 7(2):95–102

Batie SS (2008) Wicked problems and applied economics. Am J Agric Econ 90(f):1176–1191

Bertha KK (2014) Addressing the interactional challenges of moving collaborative adaptive management from theory to practice. Ecol Soc 19(1):46–53

Berkes F (1994) Co-mangement: bridging the two solitudes. Northern Perspect 22:18–20

Berkes F, Folke C (eds) (1998) Linking sociological and ecological systems: management practices and social mechanisms for building resilience. Cambridge University Press, New York

Berkes F (2007) Adaptive co-management and complexity: exploring the many faces of co-management. In: Armitage D, Berkes F, Doubleday N (eds) Adaptive co-management, collaboration, learning and multi-level governance. UBC Press, Vancouver

Beverton RJH, Holt SJ (1957) On the dynamics of exploited fish populations. Her Majesty's Stationery Office, London

Biermann F (2007) 'Earth system governance' as a croscussing theme of global change research. Glob Enviorn Chang 17:326–337

Bijlsma RM, Bots PWG, Wolters HA, Hoekstra AY (2011) An empirical analysis of stakeholders' influence on policy development: the role of uncertainty handling. Ecol Soc 16(1):51–67
Booher DE, Innes JE (2010) Governance for resilience: CALFED as a complex adaptive network for resource management. Ecol Soc 15(3):35
Bormann BT, Wagner FH, Wood G, Algeria J, Cunningham PG, Brooks MH, Friesema P, Berg J, Henshaw J (1999) Ecological stewardship: a common reference for ecosystem management. Elsevier, Amsterdam
Bruch C (2009) Adaptive water management: strengthening laws and institutions to Cope with uncertainty. In Biswas AK et al (eds) Water management in 2020 and beyond. Water Resources Development and Management. Springer, Berlin
Brunner RD, Steelman TA (2005) Towards adaptive governance. In: Brunner RD, Steelman TA, Coe-Juell L, Cromley CM, Edwards CM, Tucker DW (eds) 2005, Adaptive governance: integrating science, policy and decision making. Columbia University Press, New York
Brunner R, Lynch A (2010) Adaptive governance and climate change. Am Meteorol Soc 424
Carpenter SR, Bennett EM, Peterson GD (2006) Scenarios for ecosystem services: an overview. Ecol Soc 11(1):29–44
Carroll N (1988) Mystifying movies. Columbia University Press, New York
Carroll MS, Blatner KA, Cohn PJ, Morgan T (2007) Managing fire danger in the forest of the US inland northwest: a classic 'Wicked Problem' in public land policy. J For 105(5):239–244
Chi KS (2008) Four strategies to transform state governance. IBM Center for the Business of Government, Washington DC. Retrieved March 2010. Available at: http://businessofgovernment.org/sites/default/files?ChiReport.pdf. Accessed Mar 2010
Cole S (2001) Dare to dream: bringing the future into planning. J Am Plan Assoc 67(4):372–383
de Haan L, Zoomer A (2005) Exploring the frontier of livelihoods research. Dev Chang 36(1):27–47
Deloach JW (2000) Enterprise-wide risk management: strategies for linking risk and opportunity. Financial Times Prentice Hall, London
de Löe RC, Armitage D, Plummer R, Davidson S, Moraru L (2009) From government to governance: a state-of-the-art review of environmental governance. Final report. Prepared for Alberta environment, environmental stewardship, environmental relations. Rob de Löe Consulting Services, Guelph
Dewulf A, Francois G, Pahl-Wostl C, Taillieu T (2007) A framing approach to cross-disciplinary research collaboration: Experiences from a large-scale research project on adaptive water management. Ecol Soc 12(2):14–36
Dietz R, Ostrom E, Stern PC (2003) The struggle to govern the commons. Science 302(5652):1907–1912
Doremus H (2001) Adaptive management, the Endangered Species Act, and the institutional challenges of "new age" environmental protection. Washburn Law J 41.50
Dryzek JS (2000) Deliberative democracy and beyond: liberals, critics, contestations. Oxford University Press, New York, Oxford
Eakin H, Eriksen S, Eikeland P-O, Oyen C (2011) Public sector reform and governance for adaptation: implications of new public management for adaptive capacity in Mexico and Norway, vol 47, pp 338–351
Fabricius C, Cundill G (2014) Learning in adaptive management: insights from published practice. Ecol Soc 19(1):29–35
Farooque MA (n.d.) Dhaka Megacity: from reactive government to anticipatory governance Consortium for Science, Policy and Outcomes. Arizona State University available on the internet at: www.math.osu.edu/~tanveer/urbanconf/mafarooque.doc
Feurth LS (2009) Foresight and Anticipatory Governance. Foresight 11(4):14–32
Fishkin J (2009) When the people speak: delibertive democracy and public consultation. Oxford University Press, Oxford
Folke C (2006) Resilience: the emergence of a perspective for social-ecological systems analysis. Glob Environ Chang 16:253–267

Folke C, Pritchard L, Berkes F, Colding J, Svedin U (2007). The problem of fit between ecosystems and institutions: ten years later. Ecol Soc 12(1): Article 30 [online]

Folke C, Colding J, Berkes F (2003) Synthesis: building resilience and adaptive capacity in social-ecological systems. In: Berkes F, Colding J, Folke C (eds) Navigating social-ecological systems: building resilience for complexity and change. Cambridge University Press, Cambridge, pp 352–387

Folke C, Hahn T, Olsson P, Norberg J (2005) Adaptive governance of social-ecological systems. Annu Rev Environ Resour 30:411–473

FHSC (Foresight Horizon Scanning Center) (2009) Scenario planning. UK Government office for Science Guidance Note

Foxon R, Reed MS, Stringer LC (2009) Governing long-term social-ecological change: what can the adaptive management and transition management approaches learn from each other? Environ Policy Govern 19:3–20

Funfgeld H, McEvoy D (2011) Framing climate change adaptation in policy and practice, Working Paper 1. VCCCAR, Australia

Gunderson LH, Holling CS (eds) (2002) Panarchy: understanding transformations in human and natural systems. Island Press, Washington

Gunderson LH (1999) Resilience, flexibility and adaptive management. Conserv Ecol 3:7. Available at: http://www.consecol.org/vol3/iss1/art7. Accessed 03 Apr 2015

Guston D (2008) Innovation policy: not just a Jumbo Shrimp. Nature 454(21):940–961

Gupta J et al (2010) the adaptive capacity wheel; a method to assess the inherent characteristics of institutions to enable the adaptive capacity of society. Environ Sci Policy 13:459–471

Gupta J (2014) The history of global climate governance. Cambridge University Press, Cambridge

Gupta J, Bersgsma E, Termeer CJAM, Biesbroek GR, van den Brink M, Jong P, Klostermann JEM, Meijerink S, Nooteboom S (2015) The adaptive capacity of institutions in the spatial planning, water, agriculture, and nature sectors in the Netherlands. Mitig Adapt Strateg Glob Chang. doi:10.1007/s11027-014-9630-z

Hallegatte S (2009) Strategies to adapt to an uncertain climate change. Glob Environ Chang 19(2):240–247

Hatfield-Dodds S, Nelson R, Cook DC (2007) Adaptive governance: an introduction, and implications for public policy. A paper presented at the ANZSEE Conference, Noosa Australia, July 4, 5 2007

Henstra D (2011) the dynamics of policy change: a longitudinal analysis of emergency management in Ontario, 1950-2010. J Policy Hist 23(3):399–428

Hertin J, Berkhout F (2002) Practical experiences on policy integration and recommendations for future initiatives on EU and national level. Paper for the 3rd Bluepring workshop instruments for integrating environmental and innovation policy, Brussels, 26–27 September

Hisschemöller M, Gupta J (1999) Problem-solving through international environmental agreements: the issue of regime effectiveness. Int Polit Sci Rev 20(2):151–174

Hisschemöller M, Hoppe R (1996) Coping with intractable controversies: the case for problem structuring in policy design and analysis. Know Policy Int J Know Trans Util 8(4):40–61

Holling CS (1986) The resilience of terrestrial ecosystems, local surprise and global change. In Clark WC, Munn RE (eds) Sustainable development of the biosphere. Cambridge University Press, Cambridge, pp 292–317

Holling CS (ed) (1978) Adaptive environmental assessment and management. John Wiley, New York

Hoppe R (2011) The governance of problems, puzzling, powering and participation. The Policy Press, Bristol

Hubbard D (2009) The failure of risk management: why it's broken and how to fix it. John Wiley & Sons, Hoboken

Huitema D, Mostert E, Egas W, Moellenkamp S, Pahl-Wostl C, Yalcin R (2009) Adaptive water governance: assessing the institutional prescriptions of adaptive (co-) management from a governance perspective and defining a research agenda. Ecol Soc 14(1):26
Huntjens P et al (2012) Institutional design propositions for the governance of adaptation to climate change in the water sector. Glob Environ Chang 22:67–81
Hurlbert M, Gupta J (2015) The split ladder of participation: a diagnostic, strategic, and evaluation tool to assess when participation is necessary. Environ Sci Pol 50:100–113
Hurlbert M, Gupta J (2016) Adaptive Governance, uncertainty, and risk: Policy Framing and responses to climate change, drought and flood. Risk Analysis 36(2): 339–356
Hurlbert M, Corkal D, Diaz H, Warren J (2009) climate change and water governance in Saskatchewan, Canada. Int J Clim Chang Strat Manag 1(2):118–132
IPCC (2001) Climate change 2001: impacts, adaptation, and vulnerability technical summary, a report of working group II of the intergovernmental panel on climate change 2001, Intergovernmental Panel on Climate Change, WMO and UNEP, Geneva
IPCC (2014) Summary for policymakers. In: Edenhofer O, Pichs-Madruga R, Sokona Y, Farahani E, Kadner S, Seyboth K, Adler A, Baum I, Brunner S, Eickemeier P, Kriemann B, Savolainen J, Schlömer S, von Stechow C, Zwickel T, Minx JC (eds) Climate change 2014, mitigation of climate change. Contribution of working group III to the fifth assessment report of the intergovernmental panel on climate change. Cambridge University Press, Cambridge/New York
ISO (International Standards Office) 31000 (2009) Translated into Plain English. Available at: www.praxiom.com/iso-310000-terms.htm. Accessed 31 Mar 2015
Jones RM, Preston B (2011) Adaptation and risk management. Wiley Interdiscip Rev Clim Chang 2:296–308
Lebel L, Grothmann T, Siebenhuner B (2010) The role of social learning in adaptiveness: insights from water management. Int Environ Agreements 10:333–353
Lebel L, Anderies JM, Cambell C, Folke S, Hatfield-Dodds S, Hughes TP, Wilson J (2006) Governance and the capacity to manage resilience in regional social-ecological systems. Ecol Soc 11(1):19
Lee KN (1993) Compass and gyroscope: integrating science and politics for the environment. Island Press, Washington, DC
Lee KN, Lawrence J (1986) Adaptive management: learning from the columbia river basin fish and wildlife program. Environ Law 16:431–460
Malnes R (1995) 'Leader' and 'Entrepreneur' in international negotiations: a conceptual analysis. Eur J Int Rel 1(1):87–112
Margerum RD (2005) Collaborative growth management in metropolitan Denver: "Fig leaf or valiant effort?". Land Use Policy 22(4):373–386
McLain RJ, Lee RG (1996) Adaptive management: promises and pitfalls. Environ Manag 20:437–448
Mean E, Laugier M, Daw J, Kaatz L, Waage M (2010) Decision support planning method: incorporating climate change uncertainties into water planning. Water Utility Climate Alliance, San Francisco
Millman A, Short A (2008) Incorporating resiience into sustainability indicators: an example for the urban water sector. Glob Environ Chang 18:758–767
Mollenkamp S, Kastens B (2009) Institutional adaptation to to climate change: current status and future strategies in the Elba Basin, Germany. In: Luwig F et al (eds) Climate change adaptation in the water sector. Earthscan, London
Mostert E, Pahl-Wostl C, Rees Y, Searle B, Tàbara D, Tippet J (2007) Social learning in European river-basin management: barriers and fostering mechanisms from 10 river basins. Ecol Soc 12(1):19. Available at: http://www.ecologyandsociety.org/vol12/iss1/art19/. Accessed 3 Mar 2015
Moser CON (1998) The asset vulnerability framework: reassessing urban poverty reduction strategies. World Dev 26(1):1–19
Moser S (2009) Governance and the art of overcoming barriers to adaptation. 3:31–36

NRC (National Research Council) (2001) The drama of the commons. National Academy Press, Washington, DC

Nelson R, Kokie P, Crimp S, Martin P, Meinke H, Howden SM, de Voil P, Nidumolu U (2010) The vulnerability of Australian rural communities to climate variability and change: Part II – integrating impacts with adaptive capacity. Environ Sci Pol 13:18–27

Newig J, Günther D, Pahl-Wostl C (2010) Synapses in the network: learning in governance networks in the context of environmental management. Ecol Soc 15(4):24. [online] URL: http://www.ecologyandsociety.org/vol15/iss4/art24/. Accessed 14 Feb 2016

Nooteboom S (2006) Adaptive networks. the governance for sustainable development. Eburon, Delft

Norton BG (2005) Sustainability, a philosphy of adaptive ecosystem management. University of Chicago Press, Chicago

OECD (2012) Glossary of statistical terms. http://stats.oecd.org/glossary/. Accessed 18 May 2012

Olsson P, Gunderson LH, Carpenter SR, Ryan P, Lebel L, Folke C, Holling CS (2006) Shooting the rapids: navigating transitions to adaptive governance of socio-ecological systems. Ecol Soc 11(1):1–18

Olsson P, Folke C, Berkes F (2004) Adaptive co-management for building resilience in social-ecological systems. Environ Manag 34(1):75–90

Ormond JE (1999) Human learning, 3rd edn. Prentice Hall, Upper Saddle River

Ostrom E (2005) Understanding institutional diversity. Princeton University Press, Princeton

Ostrom E (2011) Background on the institutional analysis and development framework. Policy Stud J 39(1):7–27

Oxford Dictionaries (2010) Oxford University Press, Oxford. Available at: http://oxforddictionaries.com/definition/risk?view=uk. Accessed 8 Mar 2016

Pahl-Wostl C (2009) A conceptual framework for analyzing adaptive capacity and multi-level learning processes in resource governance regimes. Glob Environ Chang 18:354–365

Pahl Wostl C, Sendzimir J, Jeffrey P, Aerts J, Berkamp G, Cross K (2007) Managing change toward adaptive water management through social learning. Ecol Soc 12(2):30–48

Pahl-Wostl C (2007a) Transitions towards adaptive management of water facing climate and global change. Water Resour Manag 21:49–62

Pahl-Wostl C (2007b) Requirements for adaptive water management. In: Pahl-Wostl C, Kabat P, Möltgen J (eds) Adaptive and integrated water management. coping with complexity and uncertainty. Springer Verlag, Heidelberg, pp 1–22

Pahl-Wostl C (2010) Water governance in times of change. Environ Sci Pol 13:567–570

Pelling M, High C (2005) Understanding adaptation: what social capital offer assessments of adptive capacity. Glob Environ Chang 15:308–319

Pelling M, High C, Dearing J, Smith D (2008) Shadow spaces for social learning: a relational understanding of adaptive capacity to climate change within organizations. Environ Plan A 40:867–884

Plummer R (2006) Sharing the management of a river corridor: a case study of the co-management process. Soc Nat Resour 19:1–13

Plummer R (2009) The adaptive co-management process: an initial synthesis of representative models and influential variable. Ecol Soc 14(2):19. Available at: http://www.ecologyandsociety.org/vol14/iss2/art24/. Available at: http://www.ecologyandsociety.org/vol12/iss1/art19/. Accessed 3 Mar 2015

Plummer R, Armitage D (2007) Crossing boundaries, crossing scales: the evolution of environment and resource co-management. Geogr Comp 1(4):834–849

Preston BL, Brooke C, Measham TG, Smith TF, Gorddard R (2009) Igniting change in local government: lessons learned from bushfire vulnerability assessment. Mitig Adapt Strateg Glob Chang 14:251–283

Quay R (2010) Anticipatory governance, a tool for climate change adaptation. J Am Plan Assoc 76(4):496–511

Raik DB, Decker DM (2007). A multisector framework for assessing community based forest management: lessons from Madagascar. Ecol Soc 12(1):14. online http://www.ecologyandsociety.org/vol12/iss1/art14/. Accessed 3 Mar 2015

Reed MS, Evely AC, Cundill G, Fazey J, Glass J, Laing A Newig J, Parrish B, Prell C, Raymond C, Stringer LC (2006) What is social learning? Ecol Soc 15(4):19. Available at: http://www.ecologyandsociety.org/vol15/iss4/resp1/. Accessed 3 Mar 2015

Renn O (2011) The social amplification/attenuation of risk framework: application to climate change. Wiley Interdiscip Rev Clim Chang 2:154–169

Rittel H, Webber M (1973) Dilemmas in a general theory of planning. Elsevier Policy Sci 4:155–169

Rogers P, Hall AW (2003) Effective water governance, Global Water Paternership TEC Background Papers, No. 7. ISBN: 91-974012-9-3. GWP

Rowe G, Frewer LJ (2000) Public participation methods: a framework for evaluation. Sci Technol Hum Values 25(1):3–29

Sauchyn D, Pittman J, Prado S (2010) Rural vulnerability to climate change in the south Saskatchewan river basin, in the *New Normal*. Canadian Plains Research Centre Press, Regina

Scholz JT, Stiftel B (eds) (2005) Adaptive governance and water conflict. New Institutions for collaborative Planning. Resources for the Future, Washington, DC

Shannon MA, Antypas AR (1997) Open institutions: uncertainty and ambiguity in 21st-century forestry. In: Kohm KA, Franklin JF (eds) Creating a forestry for the 21st century: the science of ecosystem management. Island Press, Washington, DC, pp 437–445

Siebenhuner B (2008) Learning in international organizations in global environmental governance. Glob Environ Polit 8(4):92–116

Slovac P, Finucane ML, Peters E, MacGregor DG (2004) Risk as analysis and risk as feelings: some thoughts about affect, reason, risk and rationality. Risk Anal 24(2):311–323

Smit B, Burton I, Kleinand R, Wandel J (2000) An anatomy of adaptation to climate change and variability. Climate Change 45:223–251

Standards Australia. Australian Standard (2009) AS/NZS ISO 31000:2009. Risk management – principles and guidelines. Standards Australia, Australia

Stankey GH, Clark RN, Bormann BT (2005) Adaptive management of natural resources: theory, concepts, and management institutions, General Technical Report PNW-GTR-654. USDA Forest Service, Pacific Northwest Research Station

Swanson D, Bhadwal S (eds) (2009) Creating adaptive policies, a guide for policy-making in an uncertain world. Sage, New Dehli

Tabara JD, Zingang D, Jia G, McEvoy D, Neufeldt H, Serra A, Werners S, West JJ (2010) The climate learning ladder. A pragmatic procedure to support climate adaptation. Environ Policy Govern 20(1):1–11

Tschakert P, Dietrich KA (2010) Anticipatory learning for climate change adaptation and resilience. Ecol Soc 15(2):11–34

UNDP (2015) Inclusive development. Available at: http://www.undp.org/content/undp/en/home/ourwork/povertyreduction/focus_areas/focus_inclusive_development.html. Accessed 3 July 2015

UN/ISDR (2004) Living with risk: a global review of disaster reduction initiatives. United Nations International Strategy for Disaster Reduction Secretariat. Available at: http://www.unisdr.org/we/inform/publications/657. Accessed 3 July 2015

UN/ISDR (2005) Disaster risk reduction for sustainable development in Africa. Guidelines for mainstreaming disaster risk assessment in development. Available at: www.unisdr.org/2005/task-force/working%20groups/wg-africa/Guidelines-for-mainstreaming. Accessed 23 July 2015

Verweij M, Thomson M (2006) Clumsy solutions for a complex world, Governance, politics and plural perceptions. Palgrave Macmillan, Basingstoke

Villalba R, Grosjean M et al (2009) Long-term multi-proxy climate reconstructions and dynamics in South America (LOTRED-SA): state of the art and perspectives. Paléo 3(281):175–179
Voß J, Bornemann B (2011) The politics of reflexive governance: challenges for designing adaptive management and transition management. Ecol Soc 16(2):9. [online] URL: http://www.ecologyandsociety.org/vol16/iss2/art9/
Von Storch H (2009) Climate research and policy advice: scientific and cultural constructions of knowledge. Environ Sci Pol 12:741–747
Walters CJ (1986) Adaptive management of renewable resources. McGraw Hill, New York
Walters CJ, Holling CS (1990) Large-scale management experiments and learning by doing. Ecology 71:2060–2068
Weick KE, Sutcliffe KM (2001) Managing the unexpected: assuring high performance in an age of complexity. Jossey-Bass, San Francisco
Williams BK (2011) Adaptive management of natural resources, framework and issues. J Environ Manag 92:1346–1353
Zinn JO (ed) (2008) Social theories of risk and uncertainty, an introduction. Blackwell Publishing, Victoria

Chapter 3
Methodology: Institutional Analysis and Adaptive Governance

3.1 Introduction

This chapter sets out the methodology; it operationalizes the theoretical framework and explains how the research was conducted. This research is qualitative, deductive, emerging comparative case study research, shaped by my experience in climate change adaptation research and participation in several research projects involving Latin American and Canadian researchers (see Sect. 1.6). The methodology is a combination of the multi-level institutional method adapted from Young et al. (2005), together with four additions (see Sect. 3.2).

This chapter answers the questions: How is the theoretical framework of adaptive governance operationalized? (see Sects. 3.3 and 3.4). What are the relevant formal institutions and organizations? (see Sect. 3.3.2). What are the relevant instruments and how are they categorized (see Sect. 3.3.3)? How can the adaptive governance and policy-structuring model (Fig. 2.3) be used to analyse formal institutions, organizations, and instruments? (see Sect. 3.3.3). What are the drivers impacting the problem? (see Sect. 3.3.4). What are the instruments' effects on actors (see Sect. 3.3.5), livelihoods, (see Sect. 3.3.7) and on social learning? (see Sect. 3.3.6). How can instruments be redesigned (see Sect. 3.3.8)? Lastly this chapter interrogates the limits of the research (see Sect. 3.5) and draws inferences (see Sect. 3.6).

3.2 Evolution of the Method

This research starts with a model of multi-level institutional analysis (see Sect. 3.3.1 and Fig. 3.1) developed by Gupta et al. (2013) based on the International Human Dimensions Programme's institutional analysis framework of Young et al. (2005) and the multi-level governance theory of the Earth Systems Governance Project policy architecture (Biermann et al. 2009). Young's method focuses on

M.A. Hurlbert, *Adaptive Governance of Disaster*, Water Governance - Concepts, Methods, and Practice, DOI 10.1007/978-3-319-57801-9_3

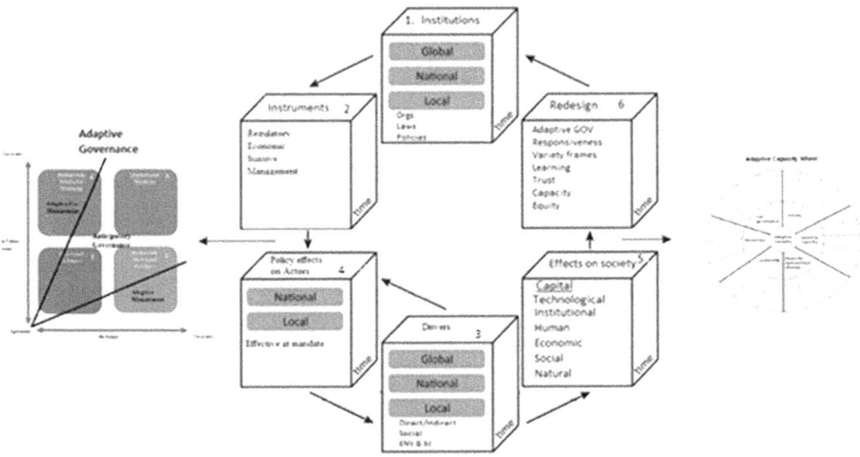

Fig. 3.1 Overview of the sequence of the multi-level institutional analysis (Adapted from Gupta et al. (2013); Young et al. (2005))

understanding the degree to which failures in governance affect a problem (the problem of causality), which instruments work and which don't in specific contexts (the problem of performance) and based on the performance analysis, how can more appropriate and effective instruments be designed (the problem of redesign). His model does not inquire into the power politics, but pragmatically focuses on what is required to improve institutions given the existing context. I have made four additions to the model:

(a) since the concept of risk is critical to disaster response d&f (Sect. 2.5.1) this element was added to the model in relation to policy structuring (see Sect. 2.4.2);
(b) in order to assess the impacts of policy instruments on actors and society, livelihood capital adaptation (see Sect. 2.5.2) has been added focusing on enhancing social, human, economic, technological, and natural capital (Moser 2009; Allison and Ellis 2001; IPCC 2001);
(c) learning has been added as part of the analysis on the policy impacts on actors and their livelihoods (see Sect. 2.4.3); and
(d) dimensions of adaptive governance (see Sect. 2.5.3) have been applied in the institutional redesign stage.

3.3 Conceptual Framework and Operationalization

3.3.1 Overview and Conceptual Model

The methodological framework used is a multi-level institutional analysis linking broader contextual variables with micro-situational variables in an attempt to understand how both social and ecological factors affect rural agricultural producers' livelihoods through adaptation and response to extreme climate events. I focus on the impact of policy via instruments (regulatory, market, suasive, and managerial), in the context of building capacity or resilience, through retention of rural producer livelihood capitals. An analytical framework has been developed to answer the overarching research question:

How can a theoretical and policy framework (norms, principles, and instruments (including regulatory, economic, suasive, and managerial)) be designed to build capacity for rural agricultural producers to respond to the increasing likelihood of d&f, defined by uncertainty?

The multi-level institutional analysis covers the following steps (see Fig. 3.1).

1. **Institutional analysis:** The current institutional framework (organizations, laws and policies) responding to d&f is identified at the global, regional, national, provincial and local level with a specific focus on the rural context;
2. **Instruments analysis:** The main instruments (regulatory, market, suasive, and managerial) emerging from the institutional framework that aim to help rural agricultural producers to respond to d&f are outlined. How the instruments and policies are framed and what environmental governance approach is used is analysed through the model of adaptive governance and the policy structuring model);
3. **Drivers' analysis:** The driving forces affecting rural agricultural producers' livelihoods and their vulnerabilities are specified. These drivers may be a) direct or indirect; at the b) global, national, or local level (Gupta et al. 2013); they are characterized as c) social (demographic, political economy), environmental, technological, or economic;
4. **Policy effects on actors:** The effectiveness of instruments at achieving their mandate, taking into the account the drivers influencing the problem, informal institutional practices, and existing independent actions taken by stakeholders is studied (see step 4). Agricultural producers are not assumed to be rational (see Sect. 1.8). The predominant social learning occurring within the case study area in response to extreme events is identified utilizing the model of adaptive governance and split ladder of participation);
5. **Effects on society – livelihood capitals:** Whether the instruments have increased the adaptive capacity of rural agricultural producers is analysed. Based on an analysis of steps 3, and 4, adaptive capacity building will be assessed by evaluating the impact of instruments on livelihood capital (human, social, economic,

technological, and natural capital) and therefore vulnerability and resilience. This will be assessed as either positive (+) or negative (−);
6. **Proposed redesign**: The policy instruments are redesigned based on what works and what doesn't work in a specific policy context; i.e. through the analysis conducted in previous steps and based on the institutional dimensions of adaptive capacity identified (see Sect. 2.5.3) (Gupta et al. 2010).

3.3.2 Formal Institutions and Organizations

Implicit in this research is the question of how to determine the most relevant formal and informal institutions (organizations, laws and policies, community livelihood strategies, etc.) and instruments on d&f in respect to rural agricultural producers and communities. This question can be further sub-divided into:

(a) How can the institutions relative to responding, or planning and adapting to d&f be identified?
(b) What are the policies, programmes, and their instruments that provide a response, or plan for and provide adaptation to d&f and how are they framed?

In order to deal with the complexity of responding to d&f, different governmental, non-governmental and private organizations are involved. Some organizations are involved in immediate response to extreme events (or disaster response), others are involved in adapting to and reducing exposure to climate variability and change; a third set are involved in planning for risks and hazards. This is partly a function of how the 'problem' of climate variability and change is framed and structured (see Sect. 2.4.2). If there is no perceived 'risk' of these events, there will be no problem for which a formal institution (government, organization, and policy) is required to respond. However, often even if a significant risk is or isn't perceived, related policies and instruments might exist which (although not framed specifically in relation to d&f) play an important role in relation to building adaptive capacity and preserving capitals in the event of an extreme event. For instance, a programme facilitating wetland restoration may have positive impacts for preventing flooding, although not framed in relation to that eventuality.

Step one of the multi-level institutional analysis involves identification of the relevant organizations for climate variability and change, climate change adaptation, and response to d&f. This answers the first secondary research question. A description of how these organizations respond to d&f will be followed by observations surrounding the power[1] dynamics inherent in these organizations and their interrelationships (obtained from interviews).

[1] Power is key in understanding the dynamics of institutions; it is also a cross-cutting theme of the ESGP informing the multi level institutional analysis (see Sect. 3.2). Power is the ability to achieve one's aims in the face of opposition from others (Egan and Chorbajian 2005). Working together with others is a form of power. It is also the capacity for acting in such a manner as to control oth-

3.3.3 *Instruments*

In each case study country, after organizations within the topic areas of adaptation to climate change and response to d&f are identified, step two of the analysis reviewed their programme and mandate through a content analysis (see Sect. 3.4.1) to ascertain relevant policy instruments.[2] Instruments can be classified into four categories: regulatory, economic or market-based, suasive, and managerial (Gupta et al. 2013: 45; Bemelmans-Videc et al. 1998; World Bank 1997). Each is defined as follows:

(a) A government or institution pursuant to its legislative function to ensure compliance implements **regulatory measures**. These are often termed command and control instruments (McManus 2009; Baldwin et al. 2011; Sterner 2003).
(b) **Economic and market** instruments either create a market or employ a market strategy. Specific behavior is encouraged through market signals rather than government directives (Stavins 2003). Private insurance as well as government disaster relief funds, and perhaps catastrophic, and pooled insurance are included (Pollner 2012). These may be backed by regulatory regimes in which case they are hybrid instruments, but predominately relate to economic compensation, which then encourages (or discourages) certain behaviours. An example in respect of drought might be a loan instrument that allowed for income smoothing through repayment provisions tied to future revenue streams (Botterill and Chapman 2009).
(c) **Voluntary, or suasive** instruments are aimed at internalizing behaviour such as increasing environmental awareness and responsibility into individual decision-making through persuasion (OECD 1994). Measures could include public voluntary programmes to encourage compliance with environmental legislation and unilateral initiatives established by groups of actors (Rivera 2002). Sometimes these voluntary programmes will have an economic benefit, but it is not the driving force in garnering cooperation. Persuasion campaigns for demand management of water (significant for reducing demand; Garrido and Gomez-Ramos 2009) could be included here as well as research and information disclosure that allows people to make more informed choices.
(d) **Managerial** instruments include those voluntary management arrangements entered into by non-government organizations (NGOs), municipalities, and civil society, and private organizations. Government could still be involved in aspects of the management scheme (Gupta et al. 2013). These are

---

ers' responses (Dahl 1956) determined by the resources possessed and issues to be decided (Held 1987). Power is non-hierarchical and competitive (*ibid.*) and may be held by one entity, or is dispersed among many (Egan and Chorbajian 2005).

[2] Policy instruments are used by state or non-state actors in order to influence behaviour and affect a certain response (Anderson 2010: 242). Policy instruments "are designed to cause people to do things, refrain from doing things, or continue doing things they would otherwise not do" (Anderson 2010: 242).

community-based arrangements, which respond directly or indirectly to events of flood or drought. An example would be local watershed organizations managing source water protection.

Table 3.1 provide examples of these instruments in the case of water, d&f. It is recognized that there is much overlap and the possibility of errors in deciding on classification. This Table is based on a review of several environmental taxonomies developed in relation to environmental governance and policy tools, but tailored to the situation of d&f. Also displayed in the Table are examples of tools and instruments for embedded problems of adaptation to climate variability and change and responding to d&f. (see Table 2.5). Each case study identifies which of these instruments was discovered within each study area.

This book focuses on property rights associated with water because of the centrality of water to agricultural production and the livelihoods of agricultural producers. Water rights can be owned privately (as a saleable market interest (existing in Alberta and Chile)), publicly (freely available to all as in Saskatchewan), or be common property (owned by the water users as in Argentina). Based on these three property rights models, the three instrument models are:

- **Regulatory:** Government agency management, generally associated with water regarded as public property; Government defers its authority for water management to an agency which assumes authority for directing who does or does not, receive water rights in accordance with bureaucratic policies and procedures. Water is owned by the State (or Crown) and interested parties are allocated water through water licences with terms and conditions.
- **Market (Economic):** Generally associated with water owned as private property as water is allocated and reallocated through private transactions. Users can trade water rights through short term, long term, temporary, or permanent transfers, reallocating rights in response to prices (Bruns and Meinzen-Dick 1995).
- **Managerial:** User-based management generally associated with water regarded as common property as water users. Those with licence or rights to water join together and coordinate their actions in managing water resources. Decision-making is collective among users. Irrigation associations are an example of this type of ownership; another example would be co-managed water resources (Plummer 2009).

At the conclusion of steps one and two of the multi-level institutional analysis, based on a content review of organizations, policies and instruments responding to climate change, d&f, an analysis of the structuring, framing, and environmental governance approaches used will be made. This analysis will utilize the model of adaptive governance and problem structuring contained in Fig. 2.3 (see Sect. 2.4.2).

Table 3.1 Classifications and description of instruments

Instrument	Description	Climate change/water management/ conservation	Drought/lack of moisture	Flood
Regulatory	Adopted by state authority	Emission reduction requirements	Licence quotas for water allocations and permits	Environmental drainage liability
	Binding	Water laws, human right to water and sanitation services	Holdback for minimum river flow	Drainage permits
	Determining what is permitted and what is not	Environmental liability, standards, bans, permits, quotas	Reclamation of unused water allocations	Flood zone building restrictions
	Includes sanctions for non-compliance	Environmental impact assessments	Water rationing	
	Without a market component	Fines for illegal water extraction		
		Land use restrictions (UNCCD 2013)		
		Emergency measures planning requirements		
		Water quality standards (Sterner 2003)		
		Waste water treatment controls, water ownership rules and rights		

(continued)

Table 3.1 (continued)

Instrument	Description	Climate change/water management/ conservation	Drought/lack of moisture	Flood
Economic/market based	Encourage behaviour through market signals	Funding adaptation and mitigation research	Crop insurance	Crop insurance
	Direct programme spending	Water tariffs, taxes or charges	Loan instruments (Botterill and Chapman 2009)	Flood insurance
	Investment in ecosystem management	Tradable permits/rights to water (Garrido and Gomez-Ramos 2009)	Grants and subsidies for new technology	Disaster relief
	Financial incentives	Payments for ecosystem services (UNCCD 2013)		Catastrophic insurance (Pollner 2012)
	Subsidies, taxes (Stavins 2003; Henstra 2015)	Export measures encouraging virtual water export		
		Subsidies, loans equity, bonds, crowd financing		

Suasive	Measures on environmental awareness and responsibility (OECD 1994)	Knowledge generation, mobilization, dissemination	Drinking water quality reports and alerts	Provision of information to public
	Public and private information, research and public awareness, demonstration by government of behaviours in government asset management and procurement practices (Henstra 2015)	Education and training	Provision of information to public	Flood alerts
		Aggregate indices	Drought prediction and alerts	
		Networks or associations	Instruments to provide access to new technology, e.g. new seeds resistant to excessive drought	
		Drinking water quality reports and alerts		
		Provision of information to public		
		Agricultural producer infrastructure grants		
		Instruments to provide access to new technology, e.g. new seeds resistant to excessive drought		
Managerial	Includes mostly self-management by private actors but could be hybrid management processes	Water attaching to land and managed by owners as common property	Insolvency	Insolvency
		Demand management of water	Drought strategy or plan	Flood strategy or plan
		Source water protection plans	Environmental best practices	Cooperation agreements for disaster recovery
		Local water governance		
		Long term water management on integrated basis		

Source: Adapted from Gupta et al. 2013: 45

3.3.4 Drivers

In step four of the multi-level institutional analysis, drivers are identified as they impact the effectiveness of instruments.

For the purposes of this research, drivers are the causes of problems or the driving forces (Gupta et al. 2013: 40) affecting rural agricultural producers' livelihoods. Examples might include changes to industrial activities, energy demand, consumption patterns, agricultural production and CO_2 emissions (Khajuria and Ravindranath 2012). Drivers may be direct (immediately impacting agricultural producers' livelihoods) or indirect (diffusely altering one or more direct drivers); occurring at the global, national, or local level (Gupta et al. 2013), characterized as demographic, economic, political, social, or natural. Examples of drivers are depicted on Table 3.2. Drivers relevant to each case study are identified in a review of secondary sources and based on perceptions of people interviewed.

3.3.5 Policy Effects on Actors

The determination of whether the instruments are effective in a specific country in the context of its given drivers is evaluated based on the effectiveness of policy instruments at achieving their mandate in relation to the actors (specifically rural agricultural producers). Effectiveness is measured in relation to the perceived ability of the instrument to achieve its mandated policy problem. This is based on a content analysis (see Sect. 3.4.1) as well as interviews (see Sect. 3.4.2). The drivers influencing the problem, informal institutional practices, and existing independent actions (individual livelihood strategies) are taken into account in this assessment.

The instruments are assessed using a criteria of (++) effective (if interviewees believed, or there was secondary literature available documenting the instrument's effectiveness), (+) moderately effective (if there was some indication from these sources of effectiveness), and (−) ineffective (if the sources established the instruments did not achieve their mandate).

3.3.6 Social Learning

This analysis includes the impact of instruments on social learning (see Sect. 2.4.3) (answering the second part of the third secondary research question). An assessment of how deep the learning has been utilizes the concepts of single, double and triple loop learning (Argyris 1999; Fabricius and Cundhill 2014) and the quadrant of the policy problem (see Fig. 2.4). Table 3.3 outlines this method. If an example of social learning (single, double, or triple) at the individual, regional or national level was discovered, an abbreviated description of the learning is placed in the table and a paragraph provided outlining this learning.

Table 3.2 Drivers impacting agricultural producers' livelihoods

Category	Driver – direct: Local	Driver – direct: National	Driver – indirect: Local	Driver – indirect: National
Demographic	Changing farm family structure	Immigration	Urbanization	Slow growing populations with low mortality and low fertility
Economic	Escalating prices of inputs	Increasing food prices	Poverty Cultural attitudes	Increasing international capital flows
Political	Reduced services in rural communities	Government strategy of neoliberalism		Market structural changes Neoliberalism/austerity
Social	Attitudes, behaviours	–	Communities maintain strong social networks Outsiders can be excluded at low cost from using resource User's support effective monitoring and rule enforcement	–
Natural	Climate change (Nelson et al. 2006) Land conversion	–	Deteriorating ecosystems and concomitant loss of services	–

Table 3.3 D&F: Framework for assessment of learning

Level/learning evidenced	**Change in practice** – single loop learning (one way information flow – Q1)	**Change in assumptions or models** – double loop learning (two way information flow – Q2 Q4)	**Change in values, norms, world views or power dynamics** – triple loop learning (iterative information flow – Q3)
Individual			
Regional			
National			

This assessment is made based on the content analysis (see Sect. 3.4.1) and interviews (Sect. 3.4.2) and my assessment utilizing the multi-level institutional analysis (step one institutional analysis, step two instrument analysis, and step four, policy effects on actors).

Table 3.4 Livelihood capitals of agricultural producers

Capitals	Description
Technological/productive/physical capital	Access to productive agricultural technology including irrigation equipment, soil erosion prevention techniques, farm machinery, etc. (Bebbington 1999)
Human	Education, skills, experience, knowledge, creativity, inventiveness, and health
Economic	The financial resources available to people (savings, supplies of credit)
Social	The quality of relations among people, or the groups, networks, norms and trust people have available to them for productive purposes (Gootaert et al. 2004)
Natural	The stock of environmentally provided assets and services such as soil, forests, minerals, water and wetlands.

Source: Based on de Haan 2000; De Haan and Zoomer 2005

3.3.7 Policy Effects on Society: Livelihoods

The policy effects on society are determined by exploring, 'What are the effects of the instruments on livelihood capital (human, social, economic, technological, and natural) in respect of d&f in the study regions?' At the societal level, these capitals represent the macro level structure of capitals which link through institutions and social relations to the micro level, or agricultural producer level (Allison and Ellis 2001).

There are several conditions that contribute to the ability of a social system to cope with stressful situations and build resilience. These conditions are defined as the 'determinants of adaptive capacity' (IPCC 2001: 895–897) that relate to the concept of vulnerability (see Sect. 2.5.2). These determinants constitute different forms of capital, or bundles of assets, which could be used in different proportions and combinations to improve the social system to reduce its risk, thereby enabling adaptation and greater resilience (Moser 2009; Moser and Satterthwaite 2008). These determinants are akin to the different 'assets' or 'capitals' that must be built to withstand extreme events of climate variability and change.

A great deal of literature identifies the various capitals necessary with much commonality, and minor differences. One source lists them as physical, financial, human, social, and natural (Moser and Satterthwaite 2008), and another as produced, natural, human, social, and cultural (Bebbington 1999). The determinants identified by the IPCC (2001) are economic wealth, technology, institutions, information and skills (human), economic wealth, infrastructure, and equity. Table 3.4 lists the capitals that will be used in this book which contain these elements (plus the addition of social and natural). Only equity is missing (which is part of both participation and inclusive development (see Sect. 2.4.3) and institutional redesign (see Sect. 2.5.3)).

Social capital is one of the important determinants of adaptive capacity. It includes social relations, formal and informal networks, group membership, knowledge sharing, values such as trust, reciprocity and civic engagement. Bonding social capital includes demographic ties to similar people; bridging social capital includes ties to people with dissimilar characteristics; while linking social capital includes ties to people in positions of authority (*ibid.*) It is a comprehensive composite of social life that promotes collective action and social cooperation (Glaeser 2001; Portes 1998). Social capital can contribute to the management of scarce resources and cope with the negative outcomes of climate variability and change by facilitating the establishment of collective objectives and organizing groups of people in pursuit of those objectives (Adger 2000); social capital can also hinder an individual's welfare, such as when group membership norms confer sharing obligations or denial of access to services (such as with respect to girls going to school) (Portes 1998).

Often 'physical' capital is identified as a determinant that could include food stocks, livestock, jewellery, equipment, tools and machinery (de Haan 2000). Given my focus on agricultural producers in each country (which include dry land farmers, irrigators, and livestock producers) access to agricultural technology is emphasized in relation to physical capital.

Human capital includes education, skills, experience, knowledge, creativity, inventiveness and health; economic capital includes financial resources available to people; and lastly natural capital refers to the stock of environmentally provided assets such as soil, minerals, wetlands, or forests.

Assessment of the influence of identified instruments on these capitals in the wake of climate change, d&f (Moser 2009; Leach et al. 1999) at the societal level was carried out through a review of secondary sources and interviews. The instruments are assessed as either (−) negatively impacting livelihood capital, or (+) positively impacting capital; the impact of missing instruments is similarly assessed in respect of (−) impact on capital.

This analyses the impacts of instruments on society. Chapter 9 will compare and contrast the case study findings to determine to what context the instruments work in different contexts.

3.3.8 Instrument Redesign

In order to redesign instruments, each country's institutions responding to climate change d&f in regards to agricultural producers will be assessed using the adaptive capacity wheel (ACW) of Gupta et al. (2010, 2014). This analysis is based on steps three, four, and five of the multi-level institutional analysis. The ACW is illustrated in Fig. 3.2 utilizing the dimensions of adaptive governance (see Sect. 2.5.3). The six dimensions of the ACW are variety, learning capacity, room for autonomous change, leadership, resources and fair governance supported by 21 criteria listed in Sect. 2.5.3.

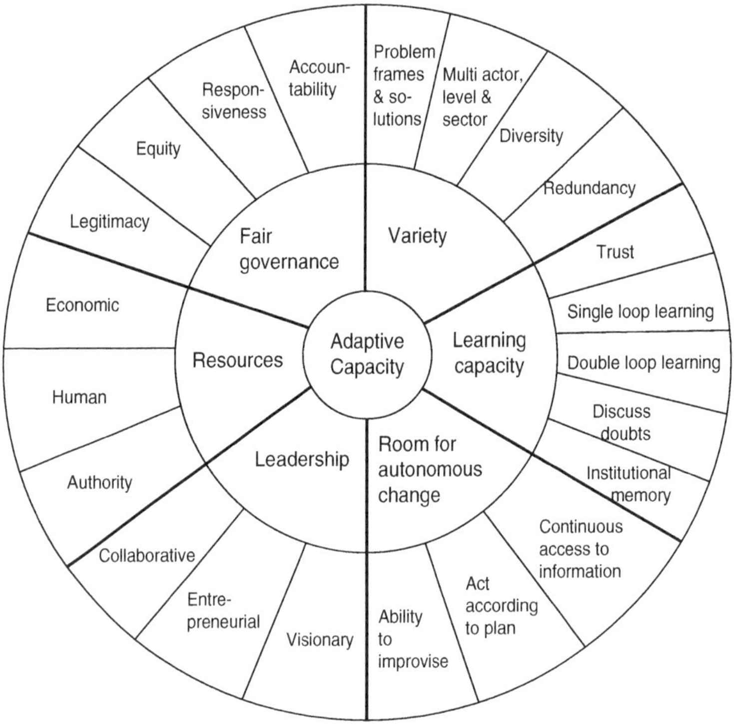

Green	Light Green	Yellow	Light Orange	Red
Very High	High	Medium	Low	Very Low

Fig. 3.2 Adaptive capacity wheel (ACW) (Source: Gupta et al. 2010)

The ACW is a qualitative assessment tool involving a normative judgement wherein the governance system responding to d&f (including formal institutions (instruments and organizations identified in each case study) and informal institutions) is ranked at an aggregate level (Gupta et al. 2015; Klostermann et al. 2010). This normative judgement is based on the content analysis (obtained in steps one and two), the analysis of problem structuring (step two), the effectiveness of instruments (step three), and the impact of instruments on capitals (step four)). The adaptive capacity dimensions of the system are ranked from very high to very low and a short paragraph given summing up the most important points for the rating. Very high green ratings reflect institutional structures enhancing adaptive capacity, light green reflect existing structures but lack of comprehensive informal institutions, yellow reflect institutions with no impact, orange reflect institutions with gaps needing to be filled and red reflect institutional structures with obstacles (Klostermann et al. 2010). No numbers are allocated in order to avoid the impression of high accuracy of the rating.

Based on this assessment, and the analysis in relation to secondary research questions two through five, recommendations will be made for policy redesign to improve the adaptive capacity of the institutions supporting agricultural producers within each case study.

3.4 Methods

3.4.1 *Content Analysis*

For each case study area, organizations (government departments, private industries, non-governmental organizations etc.) that relate to climate change, d&f and agricultural producers were identified and pertinent laws, policies, regulations, and background information were amassed as well as community responses to climate change, and extreme events of d&f. From this, relevant instruments assisting agricultural producers in responding to d&f and their stated mandate in relation to climate change, d&f were identified.

A directed deductive approach to content analysis was utilized first categorizing materials by the policy problem (climate change, d&f), and then searching for language surrounding themes of risk and uncertainty, effectiveness of instruments in relation to purpose, environmental governance approach, power of formal institutions and actors, social learning, and impact on actors and capitals (human, social, technological, economic, and natural). These themes were identified in the literature review (see 2) and methodology. This information was then supplemented by a thorough review of the previous studies, relevant pre-existing literature and research. After this, interviews were conducted exploring these themes. Interview transcripts and notes were then coded utilizing the same themes. All materials were then reviewed and analysed for major and significant findings for each theme. This analysis forms the basis of Chaps. 5, 6, 7 and 8.

3.4.2 *Interviews*

This research was able to capitalize on previous projects listed in Sect. 1.6 (354 agricultural producer interviews, 130 governance interviews in Canada and Chile surrounding climate change; 243 agricultural producer interviews and 12 governance interviews in Canada surrounding drought; and 100 agricultural producer interviews and 70 governance interviews in Canada with similar projects in Chile and Argentina surrounding d&f). Forty-one additional semi-structured interviews were conducted with key informants in the area of climate change, adaptation, d&f in each study area as detailed in Appendix I. The Field Guide for the interviews appearsin Appendix II. As well, the knowledge and work of key contacts in each country was accessed. The key contacts in Chile and Argentina identified key informants to interview (stakeholders and policy personnel) based on qualifications of expertise. Semi structured interviews were conducted exploring the themes identified in content analysis utilizing an interview guide. These interviews were coded as outlined in content analysis.

3.5 Limits

There are four main limitations of this research. First, there are limitations due to the diversity of the cases and differences inherent with cross national and cultural case study comparisons. Different contextual practices of agricultural producers in relation to the instruments and organizations studied have been explored in assessing the instruments' effectiveness and impact on livelihood capitals. Some cultural practices have been identified as drivers (see Sects. 5.3, 6.5, 7.3 and 8.3).

Second, I am Canadian and have attempted to recognize my ethnocentrism in this comparative work. However, I accept that as a Canadian, there is potential for positionality in conducting comparative case studies involving Canada.

Third, it is difficult to generalize, observe and measure agricultural producer behaviour in a particular case study. The breadth of the study reduces or obscures accuracy. As well, limits of generalization exist as often perceptions of impacts on livelihoods of d&f and influence of instruments are aggregated for all or groups of agricultural producers. The influence of other climate change impacts on agriculture (e.g. hail), elements of the agricultural sector not taken into account (e.g. supply chains), and levels of institutions were only partially considered (e.g. international). Further, the research only involves the Americas: Canada and two countries in Latin America.

Fourth, when analysing drivers and instruments, secondary sources and the perceptions of the key policy stakeholders are relied on. These perceptions and the analysis of the researcher are inherently biased. The people interviewed were not a representative sample of society, and they are not gender or culturally diverse. Data was not often available in terms of equity, gendered implications, and financial effectiveness. Thus this research is a qualitative assessment of the instruments based on impressions, rather than a quantitative analysis of the economic effects of each instrument. As a result this instrumental approach limits focus on power imbalances. Despite these shortcomings, I am convinced that the triangulation of the data has enabled robust results reflective of the situation in the case study areas.

3.6 Inferences

This chapter explained the evolution of the multi-level institutional analysis originally developed by Young et al. (2005) and added to by other scholars and further developed in this research. The theoretical framework of adaptive governance and its operationalization are detailed as they integrate with the multi-level institutional analysis. How formal organizations and instruments will be identified and their analysis utilizing the model of adaptive governance and problem structuring is explained. The assessment of social learning is outlined. Next the drivers are defined and their classification explained. Given these drivers, the analysis of policy instruments and their effects on actors is explained (utilizing the perceptions of interviewees and review of content). The instruments' effects are analysed in two ways: first a determination of effectiveness in achieving the stated objectives is made. Second, an assessment of the impact of instruments on the livelihood capitals of agricultural producers is made. Lastly, the ACW of Gupta et al. (2013) is utilized in the redesigning of instruments.

References

Adger WN (2000) Social and ecological resilience: are they related? Prog Hum Geogr 24:347–364
Allison EH, Ellis F (2001) The livelihoods approach and management of small-scale fisheries. Mar Policy 25:377–388
Anderson JE (2010) Public policymaking: an introduction. Cengage Learning, Hampshire
Argyris C (1999) On organizational learning. Blackwell, Oxford
Baldwin R, Cave M, Lodge M (2011) Understanding regulation: theory, strategy and practice, 2nd edn. Oxford University Press, Oxford
Bebbington A (1999) Capitals and capabilities: a framework for analyzing peasant viability, rural livelihoods and poverty. World Dev 27(12):2021–2044
Bemelmans-Videc M, Rist L, Vedung RC (eds) (1998) Carrots, sticks, and sermons: policy instruments and their evaluation. Transaction Publisher, Inc, New Brunswick, p 30
Biermann, F., Betwill, M.M., Gupta, J., Norichika K., Lebel, L., Liverman, D., Schropeder, H., Siebenhuner, B., with contributions from Conca, K., da Costa Ferreira, L., Bharat, D., Tay, S., Zondervan R. (2009). Earth system governance: people, places and the planet. Science and Implementation plan of the Earth system governance project. Earth System Governance Report 1. IHDP Report 20. The Earth System Governance Project, Bonn
Botterill LC, Chapman B (2009) A revenue contingent loan instrument for agricultural credit with particular reference to drought relief. Aust J Lab Econ 12(2):181–196
Bruns BR, Meinzen-Dick R (1995) Frameworks for water rights: an overview of institutional options. In: Bruns BR, Ringler C, Meinzen-Dick R (eds) Water rights reform; lessons for institutional design. International Food Policy Research Unit, Washington DC. (ISBN 0896297497) available at www.loc.gov/catdir/toc/ecip062/2005030412.html. Accessed 3 July 2015
Dahl R (1956) A preface to democratic theory. Chicago University Press, Chicago
de Haan L, Zoomer A (2005) Exploring the frontier of livelihoods research. Dev Chang 36(1):27–47
de Haan LJ (2000) Globalization, localization and sustainable livelihood. Eur Soc Rural Sociol 40(3):339–365
Egan D, Chorbajian LA (2005) Power, a critical reader. Pearson, Upper Saddle River
Fabricius C, Cundhill G (2014) Learning in adaptive management: insights from published practice. Ecol Soc 19(1):29–35
Garrido A, Gomez-Ramos A (2009) Risk management instruments supporting drought planning and policy. In: Iglesias A, Cancelliere A, White D, Garrote L, Cubillo F (eds) Coping with drought risk in agriculture and water supply systems, drought management and policy development in the Mediterranuean. Springer
Glaeser E (2001) The formation of social capital. Isuma 2(1):34–40
Gootaert C, Narayan D, Veronica NJ, Woolcock M (2004) Measuring social capital. An integrated questionnaire. World Bank Working Paper No. 18

Gupta J et al (2010) The adaptive capacity wheel; a method to assess the inherent characteristics of institutions to enable the adaptive capacity of society. Environ Sci Pol 13:459–471

Gupta J, ven der Grijp N, Kuik O (eds) (2013) Climate change, forests and REDD: lessons for institutional design. Routledge Publishers

Gupta J, Bergsma E, Termeer CJAM, Biesbroek GR, van den Brink M, Jong P, Klostermann JEM, Meijerink S, Nooteboom (2014). The adaptive capacity of institutions in the spatial planning, water, agriculture and nature sectors in the Netherlands. Mitigation and Adaptation Strategies for Global Change, pp 1–21. doi:10.1007/s11027-014-9630-z

Gupta J, Bersgsma E, Termeer CJAM, Biesbroek GR, van den Brink M, Jong P, Klostermann JEM, Meijerink S, Nooteboom S (2015) The adaptive capacity of institutions in the spatial planning, water, agriculture, and nature sectors in the Netherlands. Mitig Adapt Strategy Glob Change. doi:10.1007/s11027-014-9630-z

Held D (1987) Models of democracy. Stanford University Press, Standford

Henstra D (2015). The tools of climate adaptation policy: analysing instruments and instrument selection. Climate Policy. doi:10.1080/14693062.2015.1015946

IPCC (2001) Climate change 2001: impacts, adaptation, and vulnerability technical summary, a report of working group II of the intergovernmental panel on climate change 2001. Intergovernmental Panel on Climate Change, WMO and UNEP, Geneva

Khajuria A, Ravindranath NH (2012) Climate change vulnerability assessment: approaches dpsir framework and vulnerability index. J Earth Sci Clim Chang 3(1):1–6

Klostermann J, Bergsma E, Gupta J, Jong P (2010) Applying the adaptive capacity wheel on the background document of the content analysis. ICI2 Institutions for Adaptation, Working Document 5. IVM Institute for Environmental Studies. Amsterdam

Leach M, Mearns R, Scoones I (1999) Environmental entitlements: dynamics and institutions in community-based Natral resource management. World Dev 27(2):225–247

McManus P (2009) Environmental Regulation. Elsevier Ltd., Australia

Moser S (2009) Governance and the art of overcoming barriers to adaptation. 3:31–36

Moser C, Satterthwaite D (2008) Towards pro-poor adaptation to climate change in the urban centres of low-and middle-income countries. Climate change and cities discussion paper 3. In: Climate change and cities discussion paper 3. Washington, DC, Global Urban Research Centre, IIED

Nelson GC, Bennett E, Berhe AA, Cassman K, DeFries RT, Dobermann A, Dobson A, Janetos A, Levy M, Marco D, Nakicenovic N, O'Neill B, Norgaard R, Petschel-Held G, Ojima D, Pingali P, Watson R, Zurek M (2006) Anthropogenic drivers of ecosystem change: an overview. Ecol Soc 11(2):29. Available at: http://www.ecologyandsociety.org/vol11/iss2/art29/. Accessed 13 Apirl 2015

OECD (1994) Managing the environment: the role of economic instruments. OECD, Paris

Plummer R (2009) The adaptive co-management process: an initial synthesis of representative models and influential variables. Ecol Soc 14(2):24. Available at: url http://www.ecologyandsociety.org/vol14/iss2/art24/. Accessed 13 Apr 2015

Pollner J (2012) Financial and fiscal instruments for catastrophe risk management addressing losses from flood hazards in Central Europe. The World Bank, Washington DC

Portes A (1998) Social capital: its origins and applications in modern sociology. Annu Rev Sociol 24(1):1–24

Rivera J (2002) Assessing the voluntary environmental initiative in the developing world: the Costa Rican certification for sustainable tourism. Policy Sci 35:333–360

Stavins RN (2003) Experience with market-based environmental policy instruments. In: Mäler KG, Vincent J (eds) Handbook of environmental economics. Elsevier, Amsterdam, pp 355–435

Sterner T (2003) Policy instruments for environmental and natural resources mangement. Resources for the Future Press, co-published with World Bank and Swedish International Development cooperation Agency (Sida), Washington, DC

UNCCD (2013) Background document. The economics of desertification, land degradation and drought: methodologies and analysis for decision-making. Prepared for 2nd scientific conference, April 9–12, 2013 Bonn, Germany

World Bank (1997) Five years after Rio: innovations in environmental policy. Environmentally sustainable development studies and monograph series, no. 18. World Bank, Washington, DC, p 6

Young OR et al. (2005) Science plan. Environmental change. International Human Dimensions Programme on Global Environmental Change (IHDP), Bonn

Chapter 4
International Level

4.1 Introduction

This chapter explores the dominant institutions at the international level involved in agricultural producer livelihoods and response to climate change, d&f (see Sect. 4.4) and analyzes how they are structured and framed (Sect. 4.5). First, it examines the institutions' instruments related to climate change, adaptation, d&f, the relevant international drivers (Sects. 4.2 and 4.3), as well as social learning (see Sect. 4.6). Finally, the implications of these international organizations, instruments, and learnings are identified for redesigning instruments for agricultural producers at the case study level (see Sect. 4.7). Because of the focus on the agricultural producer of this book, a full assessment of adaptive governance of d&f at the international level is beyond the book's scope.

This chapter will illustrate that a complex multitude of international institutions have developed in the last five decades in relation to climate change, d&f. This warrants a high institutional dimension score in the ACW of adaptive governance in relation to 'variety'; there are a variety of actors, problem frames, and redundant solutions. There is also evidence of triple loop learning in relation to our understanding of pathways of resilient development and transformation. A host of suasive instruments populate the adaptive governance and problem-structuring model (see Sect. 4.4). However, there are fewer regulatory and managerial instruments; in other words learning has not yet translated into policy measures, especially at the local farm level.

International instruments don't address the global drivers of increasing trade liberalization, growing inequality, the increasing demand for energy, aging infrastructure and changing farm size. Finally, the effectiveness of instruments (specifically the UNFCCC's) is questionable as global GHG emissions continue to increase (IPCC 2014). However, the Paris Agreement of 2015 demonstrates some commitment to reducing emissions, but as the targets for countries are not included in the

M.A. Hurlbert, *Adaptive Governance of Disaster*, Water Governance - Concepts, Methods, and Practice, DOI 10.1007/978-3-319-57801-9_4

Table 4.1 International drivers

Category	Direct driver	Indirect driver
Demographic	Changing size of farms	Global population increase
		Urbanization (Levy et al. 2012)
		Growing middle class in developing world (NIC 2012)
Economic	Escalating land prices	World Recession 2008 (Verick and Islam Verick 2010; Obani and Gupta 2015)
	Changing food prices	Tempered economic development
	Changing net private capital flows	Reduction of size and budget of government
Political Economy	–	Increasing trade liberalization
		Growing inequality
		Increasing demand for energy
		Aging infrastructure
Social	–	Priority of economy over environment
Natural	Climate change	–
	Deteriorating ecosystems and concomitant loss of services (Herzog et al. 2011)	

Agreement itself, the actual effectiveness may be questionable (Gupta 2016). Further, poverty and growing inequality continue to be significant drivers.

4.2 Drivers

In order to design appropriate instruments, it is vital to understand the drivers (causes of problems or driving forces; see Sect. 3.3.4) that impact the livelihoods of rural agricultural producers, especially in relation to climate change, d&f. Hence, this section assesses the international drivers impacting agricultural producer livelihoods (see Table 4.1).

"[T]he scale, spread, and rate of change of global drivers are without precedent" (Levy et al. 2012: 4). In addition to the driver of climate change (see Sect. 1.2), there are three major themes in relation to drivers: demographic, economic, and social inequality. In relation to **demographics**, population is growing and urbanization a trend; Latin America is 75% urban, almost indistinguishable from high-income countries in this regard (Nelson et al. 2006) and birth and mortality rates are similar as well (Levy et al. 2012). Farms keep getting bigger[1] as larger farms acquire smaller

[1] Due to developments in technology (zero tillage, pest resistant crops, and rising non agricultural wages (Deininger et al. 2011:32).

ones (Deininger et al. 2011; Byerlee and Deininger 2010) and people keep moving to the city[2]; there is a growing middle class (NIC 2012). These drivers cause shortages of farm labour for agricultural producers.

Demographic trends contribute to several **economic** drivers including escalating land and water prices and changing food prices impacting agricultural producer livelihoods (NIC 2012). Between 2000 and 2012 the World Bank global food price index increased by 104.5% – an average annual rate of 6.5% (Worldwatch Institute 2013). This trend is expected to continue due to population growth,[3] increasing global affluence, and stronger linkages between agriculture and energy markets. The demand for food, water, and energy will grow by approximately 35, 40, and 50% respectively owing to an increase in global population and consumption patterns of an expanding middle class (NIC 2012: 8). The middle class is poised to expand almost everywhere in the developing world (NIC 2012: 10). In many areas of the world, agricultural land prices have hit all-time records and this trend may continue into the future (Collinson 2011; Hertel 2011). These drivers increase the income insecurity of some agricultural producers.

The increase in net private capital flows into developing countries from 1980 to 2012 has left them more vulnerable to market corrections, provoking economic contractions of the economy which is also accentuated by the liberalization of capital flows and greater financial integration (Malike et al. 2014). One important driver has been trade liberalization. Integration of the world economy has been an important strategy for many countries to promote economic growth and development. As a group, developing countries have become more important in world trade representing a higher proportion (Collier and Dollar 2002). Trade liberalization has been effectively implemented in Chile and moderately successful in Argentina, which has nationalized its pension and life insurance markets.[4]

Increasingly, the economy is ranked as important and the environmental sustainability issue is falling into second or even third place. There is continued consternation in relation to the Great Recession, which commenced in 2007. This recession is characterized by a marked global economic decline (Lightman 2009). The Great Recession has caused contracting GDPs, increasing government debt, reduction of government spending, and increasing unemployment rates in much of the United States, Europe and Canada (FRED 2013; Eurostat 2013). This driver diverts attention from the driver of climate change, issues of livelihoods of agricultural producers, and reducing inequality (Obani and Gupta 2015). There is no instrument responding to the driver of increasing trade liberalization and its impact on agricultural producers.

[2]Currently 50% of people live in urban areas; expected to climb to 60% or 4.9 billion people in 2030 (NIC 2012: 26)

[3]The global population has increased by 3.8 billion between 1961 and 2012. This is an increase of 122.9 percent (Worldwatch Institute 2013).

[4]This occurred in 2008 and has raised concerns of intergenerational equity because of a lack of long term actuarial estimates, partial coverage (not covering the informal labour force), and gender discrimination as women spend less time in the workforce (Arza 2009).

These drivers contribute to a state of growing inequality and poverty. "People living in extreme poverty and deprivation are among the most vulnerable" (Malike et al. 2014: 3). Despite poverty reduction, more than 2.2 billion people experience multidimensional poverty and 80% of the global population lack social protection (*ibid.*). Women, immigrants and indigenous people are structurally vulnerable (*ibid.*). There is a widening gap between rich and poor. In Canada, over the past three decades the Gini coefficient rose from 0.37 to 0.44 (1980–2007), representing an 18% rise in inequality (Fortin et al. 2012: 123). There are also fears about high levels of unemployment (especially among the demographically young), and frustration with the high number of people who continue to live on less than USD 2 a day (WEF 2013: 6). All of these international drivers impact the case study area and the livelihoods of agricultural prpoducer livelihoods to a varying degree.

4.3 International Institutions

4.3.1 Four Distinct Institutional Clusters and Their Evolution

Internationally, four distinct institutional clusters exist in relation to the extreme events of d&f. The first set consists of the overarching institutions that govern development, agriculture and environmental issues as well as their funding. The second set consists of governance on water issues; the third set on climate change; the fourth on response and planning for extreme events.

The 1970s were the decade of international recognition of the environment with significant developments in relation to water, climate and the recurring occurrence of disasters. The first comprehensive international conference on the environment occurred in 1972 with the Stockholm Conference on the Human Environment and the United Nations Environmental Programme was established (Johnson 2012). The oceans, the seabed and space were declared common resources of all humanity and, "the Common Heritage of Mankind was institutionalized" (Hossay 2006: 192). In 1977 the first United Nations (UN) Water Conference was held in Mar del Plata. That same decade, the World Climate Conference (WCC 1979) recognised that climate variability and change was a serious threat, but it wasn't until the late 1990s that international institutions were established in relation to it (Gupta 2014). In the same decade, the 1970s, the UN created the United Nations Disaster Relief Office (UNIDRO) staffed by a co-ordinator to promote, study, prevent, control and predict natural disasters and provide advice to government on pre-disaster planning (UNISDR 2015).

The 1980s was a decade of action, albeit with questionable effectiveness (Geary 2003). Two declarations were adopted: the UN declared the 'International Decade for Natural Disaster Reduction' (UNGA Res 43/202); and the 'International Decade of Water Supply and Sanitation' (UNGA Res 35/18). The World Commission on Environment and Development articulated the concept of sustainable development to meet, "the needs of the present without compromising the ability of future generations to meet their own needs" (WCED 1987: 43). Environmental summits and meetings continued throughout the years, but the most important success was the Montreal Protocol on Ozone Depleting Substance (Montreal Protocol 1987) coming into force in 1989 and successfully reducing halogens and chlorofluorocarbons (CFCs) (Anderson and Sarma 2002).

The 1990s until 2015 was arguably a period of contradictions. The success of the Montreal Protocol encouraged the adoption of the United Nations Framework Convention on Climate Change (UNFCCC 1992) in 1992 starting the decade of the 1990s with a bang. At the third Conference of the Parties (COP) to the Climate Convention in Kyoto Japan, in 1997, the Kyoto Protocol (KP 1997) was negotiated committing parties to internationally binding emission reduction targets, later amended in Doha in 2012 (UNFCCC 2015). However, sentiments exist that the UNFCCC has not been successful, given the continued rise of greenhouse gas emissions (Hossay 2006) (although Gupta (2014) argues that the slowdown in implementation is caused by the need for third order learning at the global level). The Paris Agreement of 2015 combines reasons to be jubilant as it is legally binding and includes a long-term objective, but is also problematic as country specific targets have been included in a separate Annex (Gupta 2016). A cause of further consternation is the 1992 Dublin Principles recognizing that water is finite, vulnerable, has economic value, and should be recognized as an economic good setting the stage for water privatization and pricing (Bakker 2007). The World Trade Organization (WTO) resulted from negotiations under the General Agreement on Tariffs and Trade (1986–1994) and hosts new trade negotiations under the 'Doha Development Agenda' launched in 2001. The WTO is a forum for negotiating, administering, and effecting trade agreements (WTO 2015). The adoption of the Sustainable Development Goals in 2015 have however prioritized food security and access to water and sanitation. These developments set the institutional stage for analysing institutions and instruments in the context of this research.

4.3.2 *Institutions*

A plethora of international institutions (organizations and instruments) exist relevant to this research. This research has focused on the most salient as of summer 2015. A comprehensive recounting is beyond the scope and breadth of this work.

The first group of international bodies function with the formal participation of States, and observer participation by civil society, NGOs, and non-profit organizations. As yet a cosmopolitan world order (wherein states are members of a single global political community (Donnelly 1994)) has not yet been created. Instead a statist order continues wherein issues of climate and climate justice are principally within sovereign national jurisdiction (*ibid.*). States participate voluntarily in these bodies, but with little legal accountability in relation to the mandates of reducing greenhouse gas emissions and vulnerability to extreme events.

The most important **intergovernmental actors and institutions** include the following:

(a) The UN is an important foundational body in relation to climate change, d&f and risk response as the various covenants and agreements of the UN General Assembly make important contributions to human rights and goal setting. In respect of disaster the UN passed a resolution to implement an International Strategy for Disaster Reduction (UNGA 2000) and created the UN Office for Disaster Reduction (UNISDR), which oversees the Hyogo Framework for Action. This framework was adopted at the World Conference on Disaster Reduction in 2005 (attended by 168 member states, 78 observer organizations, and 161 NGOs) (UNISDR 2005: 96) and endorsed by the UN General Assembly in Resolution A/RES/60/195 in 2005. In 2015, 187 UN Member States approved the Sendai Declaration and Framework for Disaster Risk Reduction 2015–2030. The 'Global Platform,' established in 2007, allows States to post their plans, identifies gaps, and tracks progress in order to accelerate national and local implementation (UNGA 2007). This public disclosure tool provides incentives for States to participate, implement, and follow the planning mechanisms provided in the Hyogo Framework.

(b) The United Nations Development Programme (UNDP) covers disaster response and has explored the possibility of a sound legal framework for expediting and providing meaningful humanitarian relief (UNDP 2004) which is not yet a reality. Future work is expected to proceed with a reaffirmation of the principle of functional interdependence between protection of individuals, their human rights, and international responsibility (Patnaik 2011).

(c) The Food and Agriculture Organization of the United Nations (FAO) covers many important themes to battle hunger, malnutrition and rural poverty including food, farming, seeds and technology, investment in agriculture, institutions, trade, and water (FAO 2015).

(d) The UNFCCC is an important framework agreement for tackling climate change. However, a legally binding agreement, which includes enforceable GHG reduction commitments for all developed countries if not developing countries, has not been achieved (Pardy 2004; FCCC/CP/2010; Gupta 2016). The Subsidiary Body for Implementation (SBI) assesses and reviews the implementation of the Convention and commitments; the Standing Committee oversees the financial mechanisms of the Convention; and the Adaptation Fund Board supervises and manages the Adaptation Fund (UNFCCC 2015). For a history of global climate change governance see Gupta (2014). The International Panel on Climate Change (IPCC) is a scientific body assessing the causes and impacts of climate change and adaptation but its recent reports have not dealt extensively with d&f (IPCC 2014).

(e) Another battery of international institutions exist in relation to water including UN Water which is the inter-agency coordination mechanism for all freshwater and sanitation related matters (UN Water) (UNWater 2015), and coordinates campaigns such as World Water Day, Sanitation for All, and the International

Decade for Action "Water for Life" 2005–2015 (UN Water 2015; Schubert and Gupta 2013; Baumgartner and Pahl-Wostl 2013).

(f) The World Bank, through its many programmes and activities has invested in reconstruction following disasters. It also has increased awareness that 'sustainable disaster reduction' must be built into the recovery process in order to reduce repetitive losses (Wisner et al. 2005: 353–354). The World Bank has two goals: "Ending extreme poverty by shrinking the share of people living on less than USD 1.25 a day to 3% by 2030, and promoting shared prosperity by raising the incomes of the poorest 40% of the population in every developing country" (World Bank 2013; 2). There are 188 members of the International Bank for Reconstruction and Development including Chile, Argentina, and Canada (IBRD 2012).

(g) The United States and other emitters have turned to less formal frameworks in respect of climate mitigation including the Major Economies Forum and the Group of Twenty, and regional plans including the Western Climate Initiative (of which Saskatchewan and Alberta are members) (CFR 2013). (However, many of these countries (including Canada) have agreed to ratify the Paris Agreement of 2015).

Many international **civil society and non-governmental** actors exist in relation to water adaptation, climate change, and response to extreme events of d&f. There are two major groupings of civil society organizations (CSO). First, disaster and humanitarian aid organizations provide services to local communities and people in times of extreme events like d&f. Second, some institutions exist for the benefit of networking and information exchange of people and entities, and sometimes advance a particular agenda (e.g. AguaFed advances the interests of the private water industry (Aqua Fed 2015)). The categories are very separate with little integration between institutions in the area of water and those of disaster response and climate change. However, within each category, there is much overlap and commonality of purpose.

Private organizations include businesses and insurance companies: reinsurers operate internationally and often companies are multinational. These organizations have their own risk management and business continuity planning and many also play a role in the response of agricultural producers and rural communities in the event of an extreme climate event. In this study, it is the insurance companies, in relation to d&f as well as the reinsurance companies, which are particularly germane.

4.3.3 Organizational Linkages and Interactions

Content analysis revealed strong institutional linkages at the intergovernmental institutional level, especially in relation to themes such as water governance, disaster, d&f. In relation to disaster most institutions coordinated with one another and participated in the UNFCCC and themes of disaster, human rights, and humanitarian relief. The UNFCCC and its COPs have the potential to bring all the organizations together to tackle all issues of climate change, adaptation to climate change, and response to d&f.

4.4 Instruments

This section classifies related international instruments (see Sect. 3.3.3) in the order of climate change mitigation, adaptation, disaster, d&f. There are no regulatory instruments specific to agricultural producers' livelihoods and response to d&f, or climate change. There are a few economic and management, but many suasive, instruments.

4.4.1 Regulatory Instruments

A handful of regulatory instruments exist with questionable effectiveness, which include those under the UNFCCC and human rights. Regarding climate change, the binding commitments of the Kyoto Protocol (an international agreement of emission reduction targets) (KP 1997) were not implemented by the US and Canada. The Doha Amendment (which set new commitments) has not yet entered into force and it is already clear that the US, Canada, Russia, and Japan will not accept targets for the period 2012–2020. The Paris Agreement (2015) includes a long-term objective to stay below 1.5–2 °C and is legally binding but the Intended Nationally Determined Contributions (INDCs) or national targets have not been included in this binding agreement (Gupta 2016).

There are three relevant human rights. First, the human right to water and sanitation services calls upon states and international organizations to provide financial resources, capacity building and technology transfer, particular to developing countries, to provide safe, clean, accessible and affordable drinking water and sanitation for all (Gupta et al. 2010). During droughts it is necessary to prioritise drinking water over other water needs and during a flood to address contamination of drinking water. Second, arguably a right to be free from damage caused by climate change exists based on legal and theoretical arguments as a rights based approach would give a "'human face' to the problem; focuses attention on generally excluded and marginalized peoples; gives the poor a voice; levels the playing field between governments and individuals, encourages accountability, good governance and promotes sustainable outcomes" (Gupta 2014:189). A potential third human right exists as some argue that there is a right to receive humanitarian aid during a disaster either as customary law (Patnaik 2011) or based on soft law instruments supported by other rights, such as the right to life (Heath 2011).[5]

4.4.2 Economic Instruments

Predominant economic instruments relate to climate change and disaster. In respect of mitigation (important to agricultural producer livelihoods as reduced emissions reduce future risk of d&f), the UNFCCC has an international emissions trading scheme. It aims to reduce emissions by certifying emission reductions (including projects funded in another qualifying country and those generated from the Clean Development Mechanism – CDM) and through allowing transfers it includes removal units based on land use, change, forestry. Although there are lawsuits and potential lawsuits surrounding the damages of emissions (Gupta 2007, 2014), this mechanism of gaining compliance has not yet been successful in bringing change in terms of reducing emissions, or the need for adaptation.[6]

Many sources of project funding exist for developing countries in relation to agricultural producer adaptation to climate change, d&f including:

(a) The UNFCCC regime operationalized the Adaptation Fund in 2008 from the share of proceeds of CDM project activities and other sources of funding in order to finance concrete adaptation projects and programmes in developing parties to the Kyoto Protocol, which include the study areas of Argentina and Chile (UNFCCC 2015). The adaptation fund provides for a sectoral focus on DRR (*ibid.*; Gupta 2014);

(b) In 1999 the World Bank established a Development Market Place grant that funds innovative, early stage development projects that are scalable and/or replicable focusing on climate adaptation and sectoral DRR (World Bank 2015). In conjunction with other banks, its multilateral "Strategic Climate Fund" (set up in 2008) provides financing to pilot new or scale up existing activities on specific climate change challenges. This fund can be accessed as either a loan or a grant and combines public sector and private sector entities (*ibid.*);

(c) The UNEP funds integrated risk and vulnerability assessments and provides training to national and local governments (UNEP 2015a). Other institutions (such as the GEF and UNDP) also provide funds for various adaptation and resilience projects;

[5] The State has the legal obligation of providing disaster assistance to its residents, and any other State offering aid must have the consent of the affected State (UN Doc. A/CN.4/629 para 78).

[6] These lawsuits are arguably an economic instrument. Small Island States have considered commencing a claim in the International Court of Justice seeking an advisory opinion in relation to the responsibilities of States under international law to ensure activities are carried out under their jurisdiction or control that emit GHG do not damage other states. This claim is based on A 4.2 of the UNFCCC and Kyoto Framework.

(d) The FAO has Government Cooperative Programme trust funds to provide technical assistance services such as policy advice to countries. It also has a TeleFood Special Fund to finance grassroots level projects to assist poor families in developing countries (FAO 2015).

The Warsaw International Mechanism for Loss and Damage Associated with Climate Change Impacts is implemented by an executive committee under the guidance of the COP of the UNFCCC. Requests are made to the SBI to consider funding irrecoverable losses and damages due to extreme events and slow onset events in relation to developing countries with common but differentiated responsibilities (Mechanism (UNCCD 2013; UNFCCC 2014; FCCC/CP/2012/8/Add.1).

Re-insurance companies provide funding to insurance companies and also utilize a risk management tool and climate impact assessment that employs a realist construction of risk (see Sect. 1.8). Reinsurance companies raise funds for these insurance purposes by issuing 'Catastrophe Bonds.'[7] The World Bank issued its first catastrophe bond in relation to sixteen Caribbean countries in 2014 linking it to natural hazard (tropical cyclone and earthquake) risks. The World Bank makes payments if no hazard occurs; in the event of a qualifying catastrophe, the insured would lose the principal (World Bank 2014).

4.4.3 *Suasive Instruments*

This section explains the suasive instruments relating to water governance, climate adaptation, and disaster. But first, an important suasive instrument in development relating to all these themes is the UNGA document on the Sustainable Development Goals adopted in 2015 which (continuing the progress of the Millennium Development Goals) (UNSDKP 2015) focuses on alleviating poverty, enhancing food security and smallholder resilience, addressing access to water, reducing vulnerability to climate change, and disaster resilience. Internationally these instruments often exist in unconnected silos but the SDG call for these Goals to be addressed coherently.

Four main instruments relate to water (with overlapping goals and mandates):

1. The Berlin Rules on Water Resources, adopted in 2004 by the International Law Association (ILA) summarize international customary law (ILA 2004). Principles include managing surface, ground and other waters conjunctively, integrating the management of water with other resources, minimizing environmental harm, maintaining ecological integrity, etc. Articles create obligations surrounding preparation for and response to floods and cooperating to mitigate drought;
2. The UN Watercourses Convention (adopted in May 21, 1997 by the UN General Assembly with both Canada and Chile sponsoring and favouring (Argentina in

[7] Bonds only have been paid once, in the event of hurricane Katrina to Zurich Financial Services by Kam Re (Kron 2008; Beder and Marshall 2011).

abstention) and entered into force May 19, 2014) provides a global framework instrument setting out rules and principles for governing international watercourses. Although it is legally binding for the ratifying states, it acts more as a suasive instrument for others as none of the case countries are parties to the Convention (although Canada and Chile are in international basins). Although Article 3(1) preserves the treaty freedom of watercourse states, the Convention encourages harmonizing existing agreements with its basic provisions (Rieu-Clarke et al. 2012). There has been a slow take up of this Convention due to an overload at the national level in implementing international agreements, low awareness and capacity, and lack of a champion like a motivated secretariat (UNWatercourses Convention 2015);

3. The United Nations Economic Commission for Europe (UNECE) Water Convention of March 17, 1992 entered into force in 1996; this Convention is open to non-European states. There are obligations to establish joint agreements and related institutional arrangements supported by an institutional framework (secretariat, working groups, etc.) for its implementation. This framework is anticipated to be operational in 2015 (UNECE 2015). As the case study countries have not yet ratified this Convention, it is suasive for them.
4. The International Network of Basin Organizations promotes integrated water resources management, defined by the Global Water Partnership (GWP) as, "a process that promotes the coordinated development and management of water, land and related resources, in order to maximize the resultant economic and social welfare in an equitable manner without compromising the sustainability of vital ecosystems" (2000: 22). A similar tool is the 'integrated land-use planning and management' tool, which is a goal or theme of the Johannesburg Plan of Implementation (WSSD 2002) paragraph 40b, Millennium Development Goal 7 (UN 2000), and UNCCD (UNCCD 1994) article 2.

International organizations have developed many instruments to manage the adaptation to climate change and response to d&f. Many other variations of these tools exist within the literature; this chart is a generalized summary. A sampling of these instruments appears in Table 4.2:

In respect of disaster, the Hyogo Framework for Action predominates, but is legally non-binding, although States generally observe its commitments (Vig and Axelrod 1999) and participate in the Global Platform of information exchange. The Sendai Declaration and Framework for Disaster Risk Reduction 2015–2030 includes seven targets (e.g. reduction in mortality, numbers of affected people, and economic losses), four priorities (understanding disaster risk, strengthening disaster risk governance to manage disaster risk, investing in DRR and resilience, and enhancing disaster preparedness for effective response) and a set of guiding principles.

Another instrument of information exchange is the UNFCCC's 'constitutive measures' published in national and regional programmes. Constitutive measures are actions that enable disaster risk reduction measures to cope with the combined effects of climate change and disaster risk (e.g. plans for protection and rehabilitation of areas affected by d&f and desertification) (Wallstrom Wallström 2009: 154;

Table 4.2 Outline of suasive instruments

Type of Instrument	Actors who utilize	Description	Risk construction
Risk management tool	Private companies – insurance and consulting	Focus on the process of managing risks (Standards Australia 2009)	Realist
DRR mainstreaming tool	States and NGOs participating in Hyogo Framework	Incorporate risk reduction practices into disaster planning to reduce loss (La Trobe and Davis 2005: 16). The goal is not just recovery, but increased resilience in the event of a future disaster and climate change (Klein et al. 2007)	Realist
Climate Impact Assessments	Researchers, NGOs, and government	A seven step model: define problem; select method; test method for sensitivity; select scenarios; assess impact on biophysical and socio-economic factors; assess autonomous adjustments; and evaluate adaptation strategies (Johnson and Weaver 2009)	Hybrid
		Both biophysical and socio-economic impacts and their interaction are integrated (Funfgeld and McEvoy 2011).	
Indicators of Vulnerability –Hyogo Framework	Some states and NGOs participating in Hyogo Framework	Indicators are to be developed in areas such as policy, strategy, geographical planning, project cycle management, external relations and institutional capacity (La Trobe and Davis 2005) measured on a litmus scale of criteria such as the degree to which the targets and indicators have been incorporated (La Trobe and Davis 2005)	Realist to hybrid
		Some literature focuses purely on government actions (UNISDR 2004) and other literature takes into account indicators of maladaptation including measures that increase GHGs, path dependency, etc. (Barnett and O'Neill 2010)	
Hyogo Global Platform	State, NGOs, disaster experts	Information and networking forum for disaster planning	Realist to hybrid

(continued)

Table 4.2 (continued)

Type of Instrument	Actors who utilize	Description	Risk construction
Hazards of place indicators of vulnerability	Researchers and NGOs	Focus is on social factors including lack of access to resources (information knowledge, and technology), limited access to political power and representation, social capital (social networks, connections, beliefs and customs) building stock and age (frail and physically limited individuals) type and density of infrastructure as well as natural factors including loss of biodiversity, wetlands etc. (Cutter 2001).	Realist
		When the indicators are correlated with hazard events and economic losses suffered, vulnerability is tested (Cutter et al. 2008)	

Article 4(1) a-j, UNFCCC). In respect of drought specifically, the UN Desertification Convention advances a Zero Net Land Degradation target (UNCCD 1994). National action programmes at the regional and sub-regional level are important mechanisms for achieving the target.

Several international organizations have specific suasive instruments. For example:

(a) UNEP has an ecosystem-based DRR (Eco-DRR) approach for adaptation and development (UNEP 2015b);
(b) FAO has a Framework for DRR in relation to the agricultural sector advancing food security through knowledge management, communication, and development of global standards particularly for vulnerable small-scale farmers (FAO 2015).
(c) The International Red Cross has developed Guidelines for international humanitarian disaster relief (IFRC 2011). Their International Disaster Response Laws, rules and principles (IDRL) Programme led to the development of the Guidelines for the Domestic Facilitation and Regulation of International Disaster Relief and Initial Recovery Assistance. This instrument regulates the common problems arising in international disaster relief operations and helps governments prepare for disasters (ICRC 2007). In 2007 these Guidelines were unanimously adopted by State parties to the Geneva Conventions[8] as well as the Red Cross and Red Crescent Movement (Giustiniani 2012). In 2008 the UN General Assembly adopted three resolutions (Res. 63/139, 63/141, and 63/137)

[8] Four treaties and three protocols of 1949 that established the standards of international law for the humanitarian treatment of wartime prisoners and civilians ratified by 196 countries.

encouraging States to make use of the Red Cross Guidelines. (These guidelines are practiced by the Red Cross in each case study region, but not yet adopted into the case study region laws).

(d) Some NGOs have established networks for exchange of knowledge including the Global Risk Forum Davos, which holds an annual International Disaster and Risk Conference (IDRC) (GRF Davos 2013). Many civil society organizations including the GWP, IWA, and WWC also perform similar functions in respect of water, and Future Earth in respect of climate change adaptation.

4.4.4 Managerial Instruments

There are few relevant international management instruments. A few bilateral and regional disaster risk management agreements have been ratified by states in order to manage disasters inter-regionally through sharing resources and knowledge and capacity building (UNILC 2007). A summary of the instruments appears in Table 4.3.

The private sector International Standards Organization (ISO) has established rules surrounding management of the environment. These standards may be enforced through procurement standards and practice (Vig and Axelrod 1999: 101). In developed countries these standards are often incorporated into standards through legislation. In most developing countries international standards are generally not legislated as regulatory requirements.

4.5 Adaptive Governance and Problem Structuring

How are these multiple and disparate instruments characterized by policy problem (see Fig. 2.3) and what environmental governance approach is utilized?

The unstructured problem of climate change is housed within the auspices of the UNFCCC and its tools and instruments.[9] The human right to water and humanitarian aid are also categorized as unstructured problems because of the competing interests and conceptions of water availability, water access and disagreements on implementation often occur. Although there is stated support for IWRM and international standards, they are often disputed in practise; conversely DRR and the principles of the Hyogo Framework or Davos Risk forum tackle flood as a wicked problem of resilience. The structured problems of d&f are met with the UNCCD, insurance, climate impact assessment and risk management. Table 4.4 illustrates that at the international level, suites of policy instruments use different management

[9]Although internationally there isn't disagreement on the science (IPCC 2014) this is still termed an unstructured problem because there is so much public and political skepticism (Baitie 2008).

Table 4.3 Summary of international instruments

Inst	Climate Change	Water	Drought/Lack of Moisture	Flood
Reg	Climate change 1.5–2 °C limit	Human right to water and sanitation services	Right to humanitarian aid	Right to humanitarian aid
Econ	Adaptation Fund	FAO TeleFood	Insurance and micro insurance	Catastrophic bonds
	Development Market Place	FAO Trust Fund	Catastrophic bonds	Humanitarian aid of ngos
	Strategic Climate Fund		Humanitarian aid of ngos	
	Emissions trading and transaction log of UNFCCC			
	Climate change lawsuits			
	Right to be free from climate change damage			
Suas	UNFCCC State Communications	Berlin Rules on Water Resources	Hyogo Platform for DRR	Hyogo Framework for Action, FAO Framework DRR
	Sustainable Development Goals	UN Watercourses Convention	Partnerships and networks (UNEP, UNFCCC, Global Risk Forum Davos, etc.)	Hyogo Platform for DRR
	Information exchange and networking of civil society and NGOs	UN ECE Water Convention	Red Cross Guidelines for humanitarian aid	Partnerships and networks (UNEP, UNFCCC, etc.)
		Platforms for Information exchange and networking of civil society and NGOs	Information exchange and networking of civil society and NGOs	Red Cross Guidelines for humanitarian aid
				Information exchange and networking of civil society and NGOs

(continued)

Table 4.3 (continued)

Inst	Climate Change	Water	Drought/Lack of Moisture	Flood
Mgt	Climate impact assessment	Integrated Water Resource Management	Hyogo Framework for Action	Hyogo Framework for Action strategic goals and seven targets
		Integrated land use planning and management	International Standards	International Standards
			Risk management tool	Risk management, DRR tools
			Indicators- Hyogo Framework – Hazards of Place	Indicators- Hyogo Framework – Hazards of Place
			World Humanitarian Summit	World Humanitarian Summit
			Zero Net Land Degradation target of UNCCD and national reporting	ISO rules

Inst Instrument, *Reg* Regulatory, *Econ* Economic, *Suas* Suasive, *Mgt* Managerial

Table 4.4 D&F: International policies

Policy	Purpose	Management approach	Problem quadrant
Climate change			
UNFCCC, subsidiary bodies and organizations combatting climate change	Reduction of GHG and adaptation to climate change	AnG	Q1-4
Rights to water and aid	Humanitarian rights implemented and measured locally	ACM	Q3,4
Climate impact assessments	Studies, initiatives, mechanisms to facilitate adaptation to climate change	AnG	Q1-4
Drought			
Convention to combat desertification	Stop land degradation	AnG	Q1-4
Water laws, networks, and conventions	International frameworks for managing water cooperatively	AnG	Q1-4
Flood			
Disaster Risk Reduction and international cooperation agreements guidelines	Disaster recovery and response	AnG	Q3

AM Adaptive Management, *ACM* Adaptive Co-Management, *AnG* Anticipatory Governance

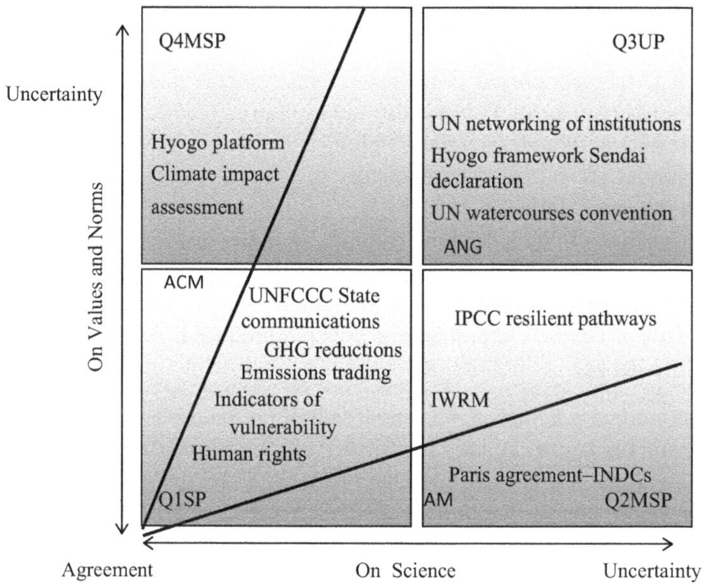

Fig. 4.1 International instruments and problem structuring

approaches to the structured problems (Q1) of d&f and unstructured problems (Q2, 3, 4; see Fig. 4.1 which shows the quadrants) of climate change.

The private sector International Standards Organization (ISO) has established rules surrounding management of the environment. These standards may be enforced through procurement standards and practice (Vig and Axelrod 1999: 101). In developed countries these standards are often incorporated into standards through legislation. In most developing countries international standards are generally not legislated as regulatory requirements.

4.6 Learning at the International Level

How do international instruments contribute to social learning? The breadth of institutions and instruments at the international level makes an assessment of learning beyond the scope of this chapter. However, three salient examples of triple loop learning (see Sect. 2.4.3) in the literature surrounding d&f emerge. These relate to resilience and the alleviation of structural social inequality, transformation, and recognition of the interconnection (or nexus) of issues such as water, energy and food. The instruments are:

(a) The Hyogo Framework has co-evolved with DRR literature. Its approach has evolved from simple objective realist risk analysis to the recognition that effective risk response includes involving people, building resilient systems, and solving issues of structural social inequality (Funfgeld and McEvoy 2011) (See Sect. 2.5.1);

(b) The IPCC has recognized in its suasive instruments that effective adaptation requires assessment of vulnerability and exposure reduction, adaptation, and transformative change in which, "climate resilient pathways for sustainable development are related fundamentally to what the world accomplishes with climate-change mitigation (high confidence)" (IPCC 2014: 28). Operationalization includes monitoring, evaluating and learning (see for example Dinshaw et al. 2014);

(c) It is now recognized that focus on the connectedness of water to energy and food using a 'nexus' sectoral approach is required (Pahl-Wostl et al. forthcoming; Hoff 2011).[10]

The case studies in Chaps. 5, 6, 7 and 8 search for the deployment of the international instruments and these learnings within the case study area (inclusive of all levels of government that have jurisdiction in the case study area). Do the local communities and their agricultural producers connect water, energy and food in a nexus? Are climate resilient pathways developed with attention to mitigation: Lastly are disasters proactively planned for with the involvement of people and attention to social inequality?

4.7 Re-designing Instruments

Analysis of the above institutions and instruments reveals multiple responses to unstructured, moderately structured, and structured problems (see Sect. 4.5). Internationally a clear linkeage is made between all issues of climate change, d&f. The policy realm of unstructured, moderately structured, and structured problems are addressed with instruments and climate change and its embedded policy problems are fully occupied, this is not so in the case study areas. However, there are few international instruments with any operational instrument that can support farmers in times of d&f. More action is needed to mobilize international suasive instruments and formalize climate commitments (Pahl-Wostl et al. 2008) into instruments implementing binding GHG reduction commitments (Gupta 2016).

There is some evidence of social learning and triple loop learning (see Sect. 4.6). This learning has occurred through the complex web of numerous interconnected institutions (see Sect. 4.3) that iteratively review instruments. Although internationally it is known that to be sustainable, development must reduce emissions. However, instruments are needed to address the driver of increasing energy demand and increasing GHG emissions. The case studies will show that international instruments and these learnings are sparsely present at the case study level and as a result

[10]This builds on the need to recognize the integrated management of water resources (see GWP 2015).

climate change, d&f may be exacerbated. Local mobilization and operationalization is required of the international third order learning (Gupta 2014)[11] (especially resilient development with reduced GHG emissions). The international level instruments are not interconnected to the local agricultural producer, perhaps through fragmented communication and implementation pathways through the various levels of government. Because of this, the international is not inclusive of the local and although this level addresses the unstructured problems of climate change, d&f, it is not inclusive of local formal and informal instituitons, agricultural producers and their livelihoods. To be adaptive, governance must interconnect the international institutions with the local institutions.

Natural disasters such as d&f continue to rise in severity and impact. Ever-increasing numbers of people (but especially the poor) suffer damage to their property, damage to their livelihoods, and loss of life (Vos et al. 2010). Contributing to this is the lack of geographical equity achieved by the current suite of international instruments (UNDP 2004: 91). This situation appears to be reinforced by the unaddressed drivers of reduced government budgets, increasing trade liberalization, increasing farm size, and urbanization. Addressing poverty and the driver of increasing global and local inequality through inclusive development appears to offer solutions to systemic problems currently surrounding access to water, vulnerability to climate change, and disaster resilience. Perhaps with attention to drivers, and focus on local impact, the international level can make the coming decade a decade of transformation. There is much room for improvement in adaptive governance focusing on agricultural producers and their communities, and one starting point begins with operationalizing the international level at the local level.

References

Anderson SO, Sarma KM (2002) Protecting the ozone layer. The United Nations history. Earthscan Press, London

AquaFed (2015) About AquaFed- Who we are. Available at: www.aquafed.org/page-7-7.html. Accessed 30 Mar 2015

Arza C (2009) BACK TO THE STATE: pension fund nationalization in Argentina. Presentation at the UNRISD Conference, "Social and Political Dimensions of the Global Crisis: Implications for Developing Countries" Geneva, 12–13 November 2009. Available at: http://www.urisd.org/80256B3C005BD6AB%2F(httpAuxPages)%2FE39AFD666528E070C12576A20056B92B%2F$file%2FArza.pdf. Accessed 3 July 2015

Baitie SS (2008) Wicked problems and applied economics. Am J Agric Econ 90(f):1176–1191

Bakker K (2007) The "Commons" versus the "Commodity": Alter-globalization, Anti-privatization and the human right to water in the Global South. Antipode 39(3):430–455

Barnett J, O'Neill S (2010) Maladaptation. Glob Environ Chang 20:211–213

[11] Special attention is paid in the case studies to determining the presence of international instruments (specifically regulatory climate change instruments and economic instruments (project funding and insurance)), and finding evidence of the influence of suasive instruments.

Baumgartner T, Pahl-Wostl C (2013) UN–Water and its role in global water governance. Ecol Soc 18(3):3. Available at: http://dx.doi.org/10.5751/ES-05564-180303. Accessed 10 Mar 2016
Beder TS, Marshall CM (eds) (2011) Financial engineering, the evolution of a profession. Wiley, Hoboken
Byerlee D, Deininger K (2010) November/December, WIDER Angle newsletter, November/December 2010, United Nations University, UNU-WIDER World Institute for Development Economics Research, Helsinki, Finland
CFR (Council on Foreign Relations) (2013) The global climate change regime. Issue Brief. Washington, DC Available at: www.cfr.org/climate-change/global-climate-change-regime/p21831. Accessed 3 July 2015
Collier P, Dollar D (2002) Globalization, growth, and poverty. A world bank policy research report. World Bank and Oxford University Press, Washington, DC
Collinson P (2011) Agricultural land prices hit record high. The Guardian, Monday 18 Apr 2011
Cutter S (ed) (2001) American hazardscapes: the regionalization of hazards and disasters. Joseph Henry Press, Washington, DC
Cutter S, Barnes L, Berry M, Burton C, Evans E, Tate E, Webb J (2008) A place-based model for understanding community resilience to natural disasters. Glob Environ Chang 18:598–606
Davos GRF (2013) Global risk forum, Home page. Available at: www.grforum.org. Accessed 25 June 2013
Deininger KBD, Lindsay J, Norton A, Selod H, Stickler M (2011) Rising global interest in farmland, can it yield sustainable and equitable benefits? World Bank, Washington, DC
Dinshaw A, Fisher S, McGray H, Neha R, Schaar J (2014) Monitoring and evaluation of climate change adaptation: methodological approaches" OECD enviornment working papaers, No. 74, OECD Publishing. Available at: http://dx.doi.org/10.1787/5jxrclr0ntik-en. Accessed 29 Mar 2015
Donnelly J (1994) Human rights and international organizations: state sovereignty and the international community. In: Kratochwil F, Manfield ED (eds) International organization: a reader. Harper-Collins, New York
Eurostat (2013) Eurostatistics, data for short-term economic analysis, issue number 06/2013, Eurstat, European Commission, Luxembourg
FAO (2015) Food and agriculture organization of the United Nations. Available at: www.fao.org/home/en/. Accessed 20 Apr 2015
Fortin N, Green DA, Lemieux T, Milligan K, Riddell WC (2012) Canadian inequality: recent developments and policy options. Canad Publ Policy 38(2):121–147
FRED (Federal Reserve Bank of St. Louis) (2013) Economic data. Available at www.research.stlouisfed.org/. Accessed 24 June 2013
Funfgeld H, McEvoy D (2011) Framing climate change adaptation in policy and practice. Working Paper 1. Melbourne, VCCCAR
Geary D (2003) Environmental movement. dictionary of American history. The Gale Group Inc. Available at: http://www.encyclopedia.com/doc/1G2-3401801397.html. Accessed 27 Mar 2015
Giustiniani FZ (2012) The works of the international law commission on 'Protection of Persons in the Event of Disasters'. A Critical Appraisal, chapter 3. In: de Guttry A et al. (eds) International disaster response law, T.M.C. ASSER PRESS, The Hague
Gupta J (2007) Legal steps outside the climate convention: litigation as a tool to address climate change. RECIEL 16(1):76–86
Gupta J (2014) The history of global climate governance. Cambridge University Press, Cambridge
Gupta J (2016) The Paris climate change agreement, China and India, Climate Law, in press
Gupta J, Ahlers R, Ahmed L (2010) The human right to water: moving towards consensus in a fragmented world. Rev Eur Comm Int Environ Law 19(3):294–305. doi:10.1080/02508069608686494
GWP (2015) The network. Retrieved from: www.gwp.org/en/About-GWP?The-network/. Accessed 30 Mar 2015

GWP (Global Water Partnership) (2000) Integrated water resources management. TAC Background Papers, No. 4. Stockholm

Heath JB (2011) Disasters, relief, and neglect: the duty to accept humanitarian assistance and the work of the international law commission. Int Law Polit 43:419–477

Hertel TW (2011) The global supply and demand for agricultural land in 2050: a perfect storm in the making? Am J Agr Econom 93(2):259–275

Herzog SK, Martinez R, Jorgensen PM, Tiessen H (Eds) (2011) Climate change and biodiversity in the tropical andes. Inter-American Institute for Global Change Research (IAI) and Scientific Committee on Problems of the Enviornment. Available at: www.iai.int/wp-content/uploads/2014/06/book.pdf. Accesssed 15 June 2015

Hoff H (2011) Understanding the Nexus. Background Paper for the Bonn 2011 Conference. The water, energy, food secuirty Nexus. Stockholm, Stockholm Environment Institute

Hossay P (2006) Unsustainable, a primer for global environmental and social justice. Zed Books Ltd, New York

IBRD (International Bank for Reconstruction and Development) (2012) Articles of agreement. As amended effective June 27, 2012. Available at: siteresources.worldbank.org/BODINT/Resources/278027-1215526322295/IBRDArticlesOfAgreement_English.pdf. Accessed 14 June 2015

ICRC (2007) Disaster response and contingency planning guide. International Federation of Red Cross and Red Crescent Societies, Geneva. Available at: http://www.ifrc.org/global/publications/idrl/resources/guidelines.asp. Accessed 14 June 2015

IFRC (2011) Introduction to the guidelines for the domestic facilitation and regulation of international disaster relief and initial recovery assistance. International Federation of Red Cross and Red Crescent Societies, Geneva

ILA (International Law Association) (2004) Berlin conference (2004) water resources law. Available at: http://www.internationalwaterlaw.org/documents/intldocs/ILA_Berlin_Rules-2004.pdf. Accessed 28 Mar 2016

IPCC (2014) Summary for policymakers. In: Edenhofer O, Pichs-Madruga R, Sokona Y, Farahani E, Kadner S, Seyboth K, Adler A, Baum I, Brunner S, Eickemeier P, Kriemann B, Savolainen J, Schlömer S, von Stechow C, Zwickel T, Minx JC (eds) Climate change 2014, mitigation of climate change. Contribution of working group III to the fifth assessment report of the intergovernmental panel on climate change. Cambridge University Press, Cambridge/New York

Johnson S (2012) UNEP The first 40 years, a narrative. UNEP, Zurich

Johnson TE, Weaver CP (2009) Environmental management. A framework for assessing climate change impacts on water and watershed systems. Environ Manag 43:118–134

Klein R, Eriksen S, Naess LO, Hammill A, Tanner T, Robledo C et al (2007) Portfolio screening to support the mainstreaming of adaptation to climate change into development assistance. Climate Change 84:23–44

Kron W (2008) Flood insurance. From Clients to Global Financial Markets. Presentation to the 4th International Symposium on Flood Defence

La Trobe S, Davis I (2005) Mainstreaming disaster risk reduction: a tool for development organisations. Tearfund, Middlesex, UK. Available at: http://www.tearfund.org/webdocs/Website/Campaigning/Policy%20and%20research/Mainstreaming%20disaster%20risk%20reduction.pdf. Accessed 13 June 2015

Levy MA, Morel AC, Adamo SB, Barr J, McMullen CP, Dieta T, Lopez-Carr D, Rosa EA (2012) Part 1: State and trends of the environment. global environmental outlook. United Nations Environment Programme, Progress Press Ltd, Valletta

Lightman D (2009) Congressional budget office compares downturn to great depression. McClatchy Washington Bureau, January 28, 2008

Malike et al (2014) Human Development Report. 2014. Sustaining human progress: reducing vulnerabilities and building resilience. UNDP, New York

Montreal protocol on substances that deplete the ozone layer (1987) A protocol to the Vienna Convention for the protection of the ozone layer. Available at: http://ozone.unep.org/Publications/MP_Handbook/Section_1.1_The_Montreal_Protocol/

Nelson GC, Bennett E, Berhe AA, Cassman K, DeFries R, Dietz T, Dobermann A, Dobson A, Janetos A, Levy M, Marco D, Nakicenovic N, O'Neill B, Norgaard R, Petschel-Held G, Ojima D, Pingali P, Watson R, Zurek M (2006) Anthropogenic drivers of ecosystem change: an Overview. Ecol Soc 11(2):29. http://www.ecologyandsociety.org/vol11/iss2/art29/. Accessed 13 Apr 2015

NIC (National Intelligence Council) (2012) Global trends 2013: alternative worlds. Officer of the Director of the National Intelligence Council, Washington, DC

Obani P, Gupta J (2015) The impact of economic recession on climate change: eight trends. Clim Dev. doi:10.1080/17565529.2015.1034226

Pahl-Wostl C, Gupta J, Petry D (2008) Governance and the global water system: a theoretical exploration. Glob Gov 14:419–435

Pahl-Wostl C, Gupta J, Bhaduri A (forthcoming) Water security handbook. Global Water Systems Project. Available at: http://www.gwsp.org/activities/current-activities/water-security/water-security-handbook.html. Accessed 12 Apr 2015

Pardy B (2004) The Kyoto Protocol: bad news for the global environment. J Environ Law Pract 144:27–43

Patnaik DS (2011) Towards an international legal framework for the protection of individuals in the event of disasters: an initial inquiry. In: Heintze HJ, Zwitter A (eds) International law and humanitarian assistance. Springer, Berlin

Rieu-Clarke A, Moynihan R, Magsig BO (2012) UN watercourses convention. user's guide. University of Dundee, Dundee

Schubert S, Gupta J (2013) Comparing global coordination mechanisms on energy, environment, and water. Ecol Soc 8(2):22

Standards Australia (2009) Australian Standard AS/NZS ISO 31000:2009. Risk management – principles and guidelines. Standards Australia

UN (2000) Millennium development goals. Available at: http://www.un.org/millenniumgoals/. Accessed 15 Sept 2014

UN Water (2015) About UN-Water. Retrieved from: www.unwater.org/about-us/en/. Accessed 13 Apr 2015

UN Watercourses Convention (2015) Online user's guide. Available at: www.unwatercoursesconvetion.org. Accessed 27 Mar 2015

UN/ISDR (2004) Living with risk: a global review of disaster reduction initiatives. United Nations International Strategy for Disaster Reduction Secretariat. 2004. Available at: http://www.unisdr.org/we/inform/publications/657

UNCCD (1994) United nations convention to combat desertification. Available at: http://www.unccd.int/en/Pages/default.aspx. Accessed Sept 2014

UNCCD (2013) Background document. The economics of desertification, land degradation and drought: methodologies and analysis for decision-making. Prepared for 2nd Scientific Conference, April 9–12, 2013 Bonn, Germany

UNDP (2004) Reducing disaster risk: a challenge for development – a global report. http://www.undp.org/cpr/disred/documents/publications/rdr/english/rdr_english.pdf. Accessed 13 Apr 2015

UNECE (2015) United nations economic commission for Europe water convention. Available at: www.unece.org/env/water. Accessed 27 Mar 2015

UNEP (2015a) The voice of the environment. Retrieved from: www.unep.org/About/. Accessed 13 Apr 2015

UNEP (2015b) Partnership for environment and disaster risk reduction. Available at: www.unep.org/disastersandconflicts/Introduction/DisasterRiskReduction/PartnershipforEnvironmentandDisasterRiskRedu/tabid/104428/Defaultl.asp. Accessed 12 Apr 2015

UNFCCC (2014) Climate change information sheet 17. Available at: http://unfccc.int/essential_background/background_publications_htmlpdf/climate_change_information_kit/items/300.php. Accessed 15 Apr 2015
UNFCCC (2015) Homepage. Available at: unfccc.int/. Accessed 30 Mar 2015
UNGA (United Nations General Assembly Resolution) (2000) Resolution 219, session 54, passed on February 2000
UNILC (United Nations International Law Commission) (2007) The protection of persons in the event of disasters, Memorandum by the Secretariat, UN doc. A/CN.4/590 (2007). para 33
UNISDR (2005) World conference on disaster reduction 18–22 January 2005, Kobe, Hyogo, Japan. In: Proceedings of the conference 'Building the resilience of nations and communities to disasters'. http://www.unisdr.org/2005/wcdr/thematic-sessions/WCDR-proceedings-of-the-Conference.pdf. Accessed 12 Apr 2015.
UNISDR (2015) History. Retrieved from: www.unisdr.org/who-we-are/history. Accessed 30 Mar 2015
UNSDKP (United Nations Sustainable Development Knowledge Platform) (2015) 2015 time for global action for people and planet. Available at: https://sustainabledevleopment.un.org/post2015. Accessed 30 Mar 2015
Verick Islam (2010) The great recession of 2008–2009: causes, consequences and policy responses. Emp/Analysis, Discussion Paper No. 4934 May 2010. Available at: ftp.iza.org/dp4934.pdf. Accessed 20 Apr 2015
Vig NJ, Axelrod RS (eds) (1999) The global environment, institutions, law, and policy. C.Q, Washington, DC
Vos F, Rodriguez J, Below R, Guha-Sapir D (2010) Annual disaster statistical review 2009: the numbers and trends. CRED, Brussels
Wallström M (2009) Disaster risk and its reduction: who is responsible? Fletcher Forum World Aff J 33(2):153–158
WCC (World Climate Conference) (1979) Geneva, 12–32 February 1979. WMO – No. 537. ISBN 92-63-10537-5
WCED (World Commission on Environment and Development) (1987) Our common future. Oxford University Press, Oxford
WEF (World Economic Forum) (2013) Global agenda outlook 2013. World Economic Forum, Geneva
Wisner B, Blaikie P, Cannon T, Davis I (2005) At risk: natural hazards, people's vulnerability and disasters. Routledge, New York
World Bank (2013) Inequality in focus. October 2013, 2(3) Available at: www.worldbank.org/content/dam/Worldbank/document/Poverty%20documents/inequality-in-focus-october2013-v12.pdf. Accessed 18 Apr 2015
World Bank (2014) Press releases. World Bank issues its first ever Catastrophe Bond Linked to Natural Hazard Risks in Sixteen Caribbean Countries. Washington, DC, June 30, 2014. Available at: http://treasury.worldbank.org/cmd/htm/FirstCatBondLinkedToNaturalHazards.html. Accessed 29 Mar 2015
World Bank (2015) World Bank – what we do. Available at: www.worldbank.org/en/about/what-we-do. Accessed 12 Apr 2015
World Watch Institute (2013) Global food prices continue to rise. Available at: vitalsigns.worldwatch.org/trends/food-agriculture Accessed June 24, 2013. posted 11 Apr 2013.
WSSD (2002) Johannesburg plan of implementation. Available at: http://www.un.org/esa/systdev/documents/WSSD_POI_PD/English/POIToc.htm. Accessed 15 Sept 2014
WTO (2015) About the WTO – a statement by former Director-General Pascal Lamy Retrieved from: www.wto.org/english/thewto_e/whatis_e/wto_dg_stat_e.htm. Accessed 30 Mar 2015

Law

FCCC, 2009/CP/2009/L.7, 18 Dec 2009
FCCC/CP, 2010/CP/2010/Cancun. Decision 1/CP.16 and Decision 1/CMP
FCCC/CP/2012/8/Add.1
KP (Kyoto Protocol) (1997) Kyoto protocol to the United Nations framework convention on climate change Available at: http://unfccc.int/resource/docs/convkp/kpeng.pdf. Accessed 19 Apr 2015
UNFCCC (1992) United framework confenction on climate change. FCCC/INFORMAL/84 GE.05-62220 (E) 200705 Available at: http://unfccc.int/resource/docs/convkp/conveng.pdf
UNFCCC (2010) Cancun agreements. COP 16/CMP 6 Cancun. Available at: nfccc.int/meetings/cancun_nov_2010/items/6005.php accessed 20 Apr 2015
UNGA (2007) Res. "International Strategy for Disaster Reduction", UN doc. A/RES/62/192 (2007), para 15

Chapter 5
Case Study Saskatchewan, Canada

5.1 Introduction

This chapter evaluates Saskatchewan's adaptive governance framework surrounding d&f. It explores how Saskatchewan's institutional policy framework (organizations and instruments) surrounding d&f impacts capitals (human, social, economic, technological, and natural) of agricultural producers given local and national drivers. Following the methodology of Chap. 3, it first describes the study area (see Sect. 5.2) and then examines the drivers impacting Saskatchewan producers (see Sect. 5.3); the pertinent organizations (see Sect. 5.4); and the relevant policy instruments (see Sect. 5.5). This allows for an examination of how policy problems are structured, risk and uncertainty are accounted for, and the nature of the environmental governance approach within the institutions of d&f governance (see Sect. 5.6). Next, an assessment is made of how effective the instruments are and the nature of learning (see Sect. 5.7), the impacts of the instruments on livelihood capitals of agricultural producers (see Sect. 5.9); and how the institutional policy framework can be characterized in relation to the dimensions of adaptive governance and hence be redesigned (see Sect. 5.10).

This chapter will demonstrate that a strong suite of effective instruments exist in Saskatchewan to respond to drought that have facilitated innovative adaptive agricultural practices and single loop learning. Strong social capital builds a human capital of adaptation knowledge and combined with effective water regulation generate resilience. A cohort of large agricultural producers access economic and technological capital including drone technology, smart seeds, and farm equipment that operates using global positioning software and real-time servicing. Economic instruments facilitate this, however many small agricultural producers can't access or choose not to access these.

Gaps exist in the institutional policy framework in relation to climate change and flood. There are no climate change or adaptation instruments at the provincial level, nor is there a federal, provincial or regional drought plan. A gap exists in climate

M.A. Hurlbert, *Adaptive Governance of Disaster*, Water Governance - Concepts, Methods, and Practice, DOI 10.1007/978-3-319-57801-9_5

change and drought with Canada's historic withdrawal from the Kyoto Protocol, the Doha Amendment and the UNCCD. This, together with the drivers of climate change, increasing global trade (with reduced government involvement), climate change science skepticism, and the absence of Q3 instruments addressing unstructured policy problems, may explain the absence of double and triple loop learning.

5.2 The Case Study Area

The South Saskatchewan River Basin (SSRB) is the largest dryland watershed in Canada (166,000 K2) (Bruneau et al. 2009). Mountains to the west impede easy access of moisture-bearing winds from the Pacific Ocean. Hence, most of the basin has a continental climate, sub humid to semiarid with short hot summers, long cold winters, low levels of precipitation (with mean annual precipitation extremely variable but generally less than 300 mm in the study area (Toth et al. 2009)). Drought response and adaptation have been a constant reality for the people of the Prairie Provinces as well as provincial, municipal, and the federal governments since settlement in the nineteenth Century. Both the Saskatchewan and Alberta study regions are located in the SSRB (Wheaton et al. 2011).

Figure 5.1 illustrates the study communities in Saskatchewan of Shaunavon and Rush Lake located in the Swift Current Creek Basin (SCCB) of the SSRB. The dominant economic activities are dryland farming followed by ranching (Bruneau et al. 2009). Around Rush Lake with a population of 65 (275 in the surrounding area) some flood irrigation is employed. The town of Shaunavon has a population of

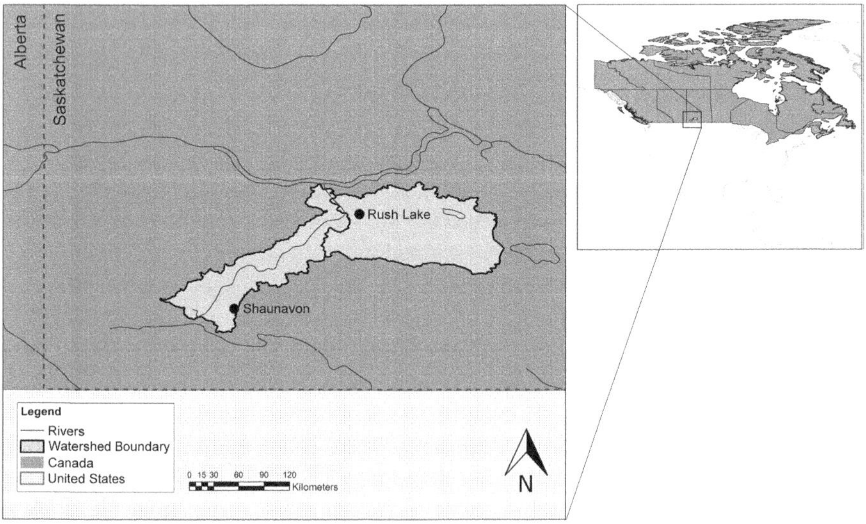

Fig. 5.1 Study regions in Saskatchewan, Canada

1691 (Statistics Canada 2012). Agriculture has been supplemented by a surge in oil and gas activity in recent years (Warren 2013) which in 2016 is on the wane.

5.3 Main Drivers Impacting Rural Agricultural Producers

In addition to international drivers (see Sect. 4.2), there are three main themes in relation to drivers: climate change, global–local economics, and the relationship between the two (see Table 5.1).

Although droughts are a common occurrence in the study area, the expected impacts of **climate change** on the basin involve more variable weather. Longer more intensive periods of drought conditions will be punctuated by periods of extreme moisture (Wheaton et al. 2016). Climate change will impact the quantity and quality of water and increase climate uncertainty. The annual mean temperature will increase between 0° and 2 °C in the 2020s and between 2 and 4 °C in the 2080s (Barrow 2010). The minimum annual precipitation will decrease between 0 and 10% in all time periods with the most severe impacts in the study region (ibid). This combined with longer, warmer winters and reduced snowpack and glaciers in the mountains is already changing stream flow (Byrne et al. 2010). The 2014 IPCC report predicts a warming atmosphere which will hold more water; the warmer artic, weaker jet stream, and weaker winds from west to east will stall weather patterns making conditions right for summer storms lasting for days (which previously lasted only hours) (IPCC 2014).

The **global-local economic** drivers interrelate. Increasing trade liberalization and reduction in government programmes and instruments is exposing agricultural producers to more uncertainty. Government austerity has resulted in: (1) a reduction in program funding; (2) a transition from programmes subsidizing agricultural pro ducers into self-funding programmes; and (3) an absence of progressivity in programmes (such that small agricultural producers receive no greater benefit than large).[1] Agricultural extension services are ending as well. In 2012 a longstanding Canadian agricultural institution, the Canadian Wheat Board (CWB) (historically the global vender of Canadian wheat) lost its single desk status for wheat sales. Producers are now free to ship their grain across the border into the United States themselves and sell to any purchaser. Whereas the CWB had returned all profit to the farmers for this service, private companies now will not.

Saskatchewan producers are increasingly exposed to global market prices and there is a concurrent trend to higher costs and higher debt (Statistics Canada 2011) that is resulting in a tighter squeeze between high input costs and low commodity prices. Increasingly, large multinational corporations dominate multiple links in the

[1]The current federal farm stabilization programmes total 3 billion in payments over five years; payments for the three years (1991–1994) totalled 4.5 billion and these had been reduced since the 1980s (Office of the Auditor General of Canada 1994).

Table 5.1 Drivers impacting agricultural producers' capitals

Category	Driver – direct		Driver – indirect	
	Local	National	Local	National
Demographic	Increasing size of farms	–	Urbanization/population growth	–
	Shortage of farm labour (Wandel et al. 2010)		Aging producers	
Economic	Escalating land prices (Remax 2014)	–	Reduction of size and budget of government	World recession 2008
				Tempered economic development
				Reduction of size and budget of government
Political Economy	Increasing cost of inputs	Reduction of government services and programs	Aging infrastructure (roads, dams)	Increasing trade liberalization
		Larger more technologically advanced farm equipment raising costs resulting in increasing farm debt (Statistics Canada 2011)		Growing inequality (Yalnizyn Yalnizyan 2013)
				Increasing demand for energy
				Aging infrastructure
Social	Climate scepticism	–	–	Priority of economy over environment
Natural	Climate change	–	Deteriorating ecosystems and concomitant loss of services (FPTGC 2010)	–

food chain.[2] Producers are getting older[3] with an increasingly larger agricultural unit and a shortage of labour.[4]

[2] e.g. 95% of beef processing in Canada is done by two major corporations: Cargil and JBS Food Canada, a subsidiary of a Brazil company (Graveland 2013).

[3] Nearly half of the farmers are 55 years of age or older (Statistics Canada 2012; Johnstone 2012).

[4] Further, five % of farms generate nearly half of total farm cash receipts (Johnstone 2012). The number of farms have declined by 10.3% nationally and 16.6% in Saskatchewan between 2006 and 2011 (a trend since 1941) (Statistics Canada 2012; Johnstone 2012).

The Relationship Between Economics and Climate Change Is Strained The previous Prime Minister of Canada announced that he doesn't believe any country would support a climate change agenda at the expense of that nation's economy (Wyld 2014). Also troubling is the Canadian federal government's significant funding cuts to science and muzzling of scientists by preventing them from releasing results and speaking to the media (Hill 2014; CAUT 2013; C1). This situation is embedded in a Canadian public whose third greatest concern is, "A severely environmentally degraded future" (Ekos 2013: n.p.). This concern is a documented reality (FPTGC 2010) and emphasized by a Saskatchewan auditor's warning that more money should be invested in the environment to avoid contamination of water and soil (Ferguson 2013).

5.4 Institutions (Organizations) That Build Capacity for Climate Change, d&f

Canada is a federal constitutional democracy. Both federal and provincial governments have significant roles in relation to climate change, d&f, although the province predominates. Local governments are responsible for response to emergencies such as floods. Institutions and instruments will be described in relation to water, climate change, and then d&f.

In Canada, because of the division of power between the provinces and the federal government, the province is the predominant level of government involved in **water** and supports municipalities in respect of d&f (Hurlbert 2009). Although the Saskatchewan Water Security Agency (WSA) dominates water governance, when an event of d&f occurs, the municipality is the first responder, and if several municipalities become involved, the Ministry of Government Relations manages the event by relying on the advice of the intergovernmental Drought and Excess Moisture committee. Thus all government ministries are coordinated through the committee's work (GS 2012). Municipalities and irrigation districts are currently tasked with maintaining infrastructure, preparing emergency response plans for floods, and providing a first response in the event of a flood. These provincial institutions coordinate with national Public Safety Canada (which ultimately provides disaster assistance funds).

Natural Resources Canada (NRCan) heads **climate change** research and adaptation for the federal government and Environment Canada is tasked with water and climate research. Institutions promoting adaptation and building capacity of producers include Ag Canada, NRCan and the Ministry of Agricultural for Saskatchewan. These government departments develop and maintain programmes that build resilience and assist agricultural producers in stabilizing farm income during periods of instability, or weather variability including d&f.

Many national and provincial civil society and non-profit organizations exist with mandates conducive to adaptation to climate change, d&f. However, at the provincial and local level these organizations have very focused mandates with little interconnection, generally focusing directly on water or environmental issues. A further group of institutions exist in relation to responding to extreme events of flood. The only exception to this silo characteristic is PARC (an academic institution working on climate adaptation) and the local watershed organizations involved in programmes assisting agricultural producer adaptation; these organizations have participated in adaptation planning (Rowan et al. 2011).

The role of CSOs and NGOs in relation to d&f has increased in the last decade due to the withdrawal of services by the Canadian government. The Red Crescent, Red Cross Society and the Salvation Army are active in responding to disasters. In Saskatchewan the Red Cross is contracted to provide counseling post disaster (flood) (S1). The development of local institutions at the community level, such as Southwest Search and Rescue and the Southwest Public Safety Region Inc., also fill a void left by the retreat of the federal government in providing planning assistance and emergency training to local communities because of the closing of the Canadian Institute of Emergency Planning (A1).

Private organizations include insurance brokers, and a growing number of agricultural producers are evolving into (incorporated) agri-businesses and intensive livestock operations. Oil extraction and related businesses are also growing (C11).

Interconnecting agencies include the Prairie Provinces Water Board (PPWB) (constituted by the Federal and Provincial governments (represented by Alberta Environment and Sustainable Resource Development (AESRD), WSA, Environment Canada, and Agriculture Canada) that manages interconnecting watercourses and watershed associations; irrigation districts also play a local coordinating role in relation to water. Irrigation districts obtain a licence of water and then internally govern sub-licences within the district.

5.5 Instruments Responding to Climate Change, d&f

The main instruments (see Sect. 3.3.3) that assist agricultural producers (see Table 5.4) are identified and grouped by function on water, climate change, d&f. (Because the federal government financially supports these instruments, many of these instruments are also available in Alberta).

5.5.1 Regulatory Instruments

The Saskatchewan WSA oversees a very robust regulatory water governance system that does not include an economic instrument (as in Alberta). It is legislatively empowered to manage water economically and efficiently (WSA Act) in a technocratic manner in Q1 with minimal adaptive co-management (see watershed committees below in 'management'). The Crown (provincial government) owns all water.

Table 5.2 Institutional water instruments

Instrument	Description
Principle under which water is managed	Common property – belonging jointly to Saskatchewan people
Allocation of water rights	Licensed interests allocated by the Water Security Agency on conditions considered appropriate
Priority of use	No statutory priority scheme
Water Market	None
Water allocation dispute resolution	Water Appeals Board – internal government entity
Potable water accountability	Local level – regulatory drinking water standards and required reporting
Governance Accountability	A Provincial Crown Corporation (WSA) is vested with management of water Its board reports to a Provincial Crown Investment Corporation of the Government
Water price	Tariff set by municipal water supplier for water and sanitation services

Source: WSA Act; Hurlbert 2009

Interests in water are issued by licence with specific terms and conditions overseen by the WSA (WSA Act: see Table 5.2). There is no first in time, first in right priority scheme, or other priority scheme (governing community drinking water versus irrigation priority, for instance) (Hurlbert 2009).

In Saskatchewan, there are few regulatory instruments responding to climate change, d&f. Regulations pursuant to the federal Clean Air Act reduce car and industry emissions. In July 2015 restrictions on the amount of emissions of coal-fired power generation came into effect (EC 2012; CEPA 1999). No government assessments of these regulatory measures are made in relation to overall GHG reductions. Therefore, I conclude that these initiatives reduce GHG in a Q1 structured manner, with no engagement of adaptive management processes, whereby hypothesis testing and measurement of reduction targets occurs.

Canada withdrew from the Kyoto Protocol in 2011 (CBC 2011) (but still participates in the UNFCCC COPs). However, the Prime Minister has recently communicated to the UNFCCC an economy wide target of reducing annual emissions to 30% below 2005 levels by 2030 (Canada 2015) and has advised that Canada will ratify the Paris agreement.

Canada withdrew from the UNCCD in 2013 (Hamadijan 2013) although previously stressing its importance due to lands (including the case study areas) "at risk of desertification" (Canada 2002: 7). (Alarmingly, at the same time, **drought** has been taken off of Canada's disaster database). These two withdrawals leave a gap of international institutions applicable to the multi-level analysis of Canada.

Since 1989 emergency responses to **floods**, have been governed through legislation setting out the powers and responsibilities of municipal government and providing for a legislated scheme of provincial and federal disaster assistance financing (Henstra 2011). The latter will be discussed as an economic instrument.

Saskatchewan's *Emergency Planning Act*, S.S., c. E-8.1 obligates every municipality to plan for and be responsible in the case of an emergency (which includes a calamity caused by forces of nature) unless the Province declares a state of emergency and takes control. Pertinent flood regulations include: drainage permits, fines for illegal drainage, civil court claims for damages from wrongful drainage, requirement that Saskatchewan's municipal infrastructure be built to a one in 500 standard (i.e. that an event will occur in one year with a probability of .002) (S2), building code requirements responding to flood, and flood zone building restrictions. The Canadian state does not have measures to: (1) restrict building in flood plains, (2) ensure the availability of current flood risk maps; and (3) differentiate between more than one or two levels of flood risk (Sandink et al. 2010).

5.5.2 Economic Instruments

There is a deficit of economic instruments responding to climate change, but a robust response to d&f. Currently, there are no federally funded initiatives for climate **adaptation** in the study area. Other than a few initiatives such as contributing funding for carbon capture and sequestration (CCS) and cleaning coal (while burning it in a coal plant) there are no **mitigation** measures.

Although not framed specifically in relation to **adaptation**, many farm programmes exist (entitled "Growing Forward" – AgriInvest, AgriStability, AgriInsurance, AgriRecovery) and provide government loans (C11), which alleviate changes in farm income that may have resulted from d&f, or such things as changing commodity prices (GC 2012; GS 2013b). These economic instruments generally reside in Q1. New programmes have been created to respond to the driver of trade liberalization by making agricultural producers more competitive by enhancing innovation, marketing and enhancing competitiveness (Agri-innovation) (GC 2012; GS 2013b).

In 1935 the federal government established rural water programming to address **drought** following the devastating multi-year droughts in the 1920s and 1930s. The Saskatchewan Farm and Ranch Water Infrastructure Programme (FRWIP) continues this type of programming in order to expand the livestock industry, encourage rural economic activity, and mitigate the effects of future drought (all Q1 issues). Projects such as community wells, large and small diameter wells, shallow or deep buried pipelines and dugouts are eligible for funding. Project costs are shared between the proponent (i.e., producer or municipality) and the federal and provincial governments (GS 2011).

In respect of disasters (**flooding**) the Saskatchewan government has a Provincial Disaster Assistance Programme (and receives funding from the federal government Disaster Financial Assistance Arrangements Act (GC 2013a; PSC 2014)). This programme provides financial compensation for pre-disaster value of certain property lost in a declared emergency (GS 2013a) thereby responding to flood emergency in a reactionary, Q1 manner with little, if any, public involvement in planning. Flood insurance is not available.

5.5.3 *Suasive Instruments*

The 25 Year Saskatchewan Water Security Plan incorporates climate change research and adaptation (SWSA 2012). Many suasive and information measures exist in Saskatchewan helping rural agricultural producers adapt to d&f, but are not structured in relation to climate adaptation. The Canada-Saskatchewan, and Canada-Alberta Farm Stewardship Programme (FSP) assist producers in **adapting** to water shortages by helping them in writing environmental farm plans (GS 2015) to respond to environmental risk and water supply threats, thereby potentially reducing producers' vulnerability to climate and environmental change (S3, Hurlbert and Pittman 2014). Information relating to **drought** is provided through the 'Drought Watch' website (AAFC 2015).

In 2009 Canada established a Platform for Disaster Risk Reduction as part of its commitment to the UN's International Strategy for Disaster Reduction's Hyogo Framework for Action. The Canadian government instituted a National Disaster Mitigation Strategy (including all levels of government) in 2013. Primary actions have yet to be identified by a committee (GC 2013b). This platform is an assembly of interdisciplinary stakeholders brought together by their shared interest in DRR (GC 2013c). The federal and Saskatchewan governments are sources of online information for municipalities and entities undertaking emergency management planning. However the Canadian Emergency Management College (which trained people in emergency management) was closed in 2012 and now provinces are solely responsible for training municipal employees. Increasingly active, the Red Cross responds by implementing rights and guidelines to humanitarian aid.

These measures indicate the existence of Q4 tools (the Platform for Disaster Risk Reduction) and Q1 tools (drought information). Although the FSP assists in adaptation to d&f, it is framed only in relation to d&f and not the unstructured Q3 problem of climate change.

5.5.4 *Managerial Instruments*

Significant management instruments exist in relation to water governance and adaptation. However, there is a deficit in relation to d&f. In 2002 Saskatchewan's intergovernmental **drought** monitoring committee drafted a Drought Risk Management Plan for Saskatchewan, but this has never been finalized (AAFC 2002). No regional or provincial flood plans exist. In respect of disaster the government contracts services of counselling, shelter and food with the Red Cross.

There are two significant instruments relating to water governance: (1) The Boundary Waters Treaty of 1909 (International Boundary Waters Treaty Act, R.S.C., 1985, c. I-17) apportions inter-basin transfers of water between Alberta, Saskatchewan, and Manitoba. This agreement does not deal with excess water, or drainage. The agreement protects downstream Saskatchewan by ensuring 50% of SSRB flows coming from Alberta.

(2) Watershed groups (whose members are predominantly urban and rural representatives) have commenced planning for **adaptation to d&f.** It is a Q3 example of tackling moderately structured issues affecting water quality and quantity through iterative information exchange and citizen participation. The North Saskatchewan River Watershed (Rowan et al. 2011), and the Upper Souris River Watershed (East et al. 2012) each have developed plans. The provincial WSA and NRCan (S3) facilitated the plans.

An elaborate system of insolvency management exists for agricultural producers in Canada situating the results of d&f and drivers (see Sect. 5.6) in Q1. When a producer is unable to meet its financial obligations, a suite of laws and regulations assist in initially providing a period of up to 1 or 2 years to make the necessary payments, and thereafter the ability to declare bankruptcy. The declaration of bankruptcy allows the producer to serve a period of bankruptcy (of several months or years) and eventually be discharged from bankruptcy free of most, if not all, debt obligations. Certain exemptions allow agricultural producers to retain their home quarter upon which they live and certain farm machinery and tools. Often the producers are able to lease their land back from the financial institution for a period of years and then have a right of first refusal to purchase this land when sold by their financial institution (C11).

5.6 Adaptive Governance and Problem Structuring

Saskatchewan instruments are mostly framed in a Q1 manner. Some instruments show promise for dealing with Q3 unstructured problems with increasing public engagement. Figure 5.2 summarizes the instruments responding to d&f by policy problem and environmental governance approach.

Figure 5.2 shows that the unstructured Q3 policy problems of climate change mitigation and adaptation do not have a consistently accessible policy instrument such as a carbon tax or market. A few regulatory measures exist that are managed in a Q1 manner without monitoring of their actual contribution to GHG reduction. The exception is watershed groups that occasionally develop plans to respond to unstructured problems of d&f in Q3 and moderately structured problems of source water protection in Q4. Many policy instruments exist in Q1 responding to structured embedded policy problems of d&f (with flood responded to as an emergency). A suite of economic instruments assist producers with the structured problem of responding to events of d&f (Q1); the FSP programme with its environmental farm plans assists in measures which build resilience to d&f.

Some movement has occurred between quadrants. The FRWIP is managed in Q1, but in 2008 local producers created this programme and managed it with government support in Q4 (adaptive co-management); it wasn't until year three that the government took over its management placing it in Q1. The examples of Q3 outlined (e.g. Watershed drought plans and the Water Security Plan) demonstrate occasional engagement with the public in relation to wicked problems and a migration of out-

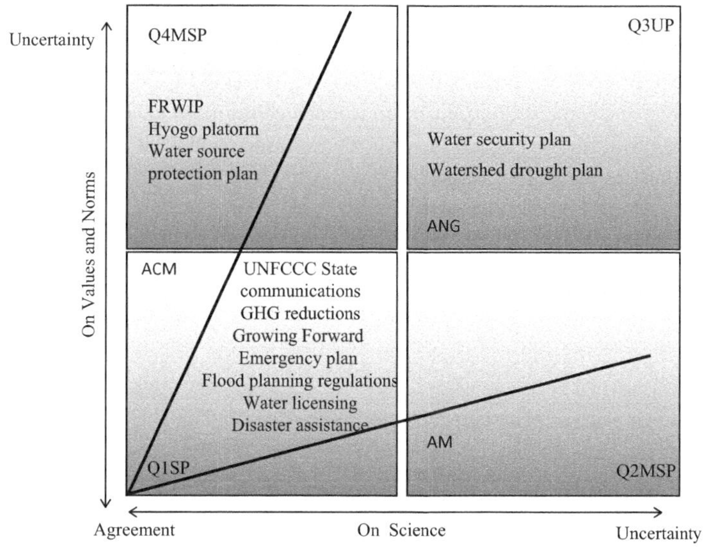

Fig. 5.2 Saskatchewan instruments and problem structuring
ACM (Adaptive co-management), Q4ACM (Quadrant 4 moderately structured problem), ANG (Anticipatory governance), Q3UP (Quadrant 3 unstructured problem), Q2MSP (Quadrant 2 moderately structured problem), AM (Adaptive management), Q1SP (Quadrant 1 structured problem)

comes into either Q4 issues (source water protection) or Q1 (provisions of information for d&f). No evidence of adaptive management or effective interrogation of uncertainty of science in Q2 was discovered.

5.7 Impacts of Instruments on Actors Measured by Mandate Effectiveness

The impact, given the drivers (see Sect. 3.3.5 and Table 5.4) is assessed in relation to instrument effectiveness, by instrument function.

5.7.1 Regulatory Instruments

Although Saskatchewan has had many **droughts**, particularly in the study area, the government's regulatory licensing of water has responded to the shortage, and no producer or community has been without water. Often irrigators receive less water and communities are asked to reduce their water use. These regulatory actions have solved the shortage problems (Warren 2013; S2) without resort to economic instruments (such as increasing the price of water).

Other regulatory measures haven't been as effective. To date **climate mitigation** measures have been ineffective; statistics of the UNFCCC report that Canada had a 49% increase in GHG emissions from 1990 to 2011 (UNFCCC 2013; GC 2014).

Regulatory instruments pertaining to **flood** are fragmented. Drainage permits or fines for illegal drainage (flood) haven't worked due to lack of personnel to implement (Hurlbert et al. 2015). Regulatory measures consisting of building standards (codes and prevention of building in flood zones) do not account for climate change (S2, S4). They are set at the national level and involve a very political process (C5). Sandink et al. (Sandink et al. 2010) conclude that the Canadian government needs to clarify how people in high-risk flood zones can be moved.

5.7.2 Economic Instruments

Economic instruments have been successful in building the adaptive capacity of rural agricultural producers, for **droughts** up to 2 years (Diaz and Warren 2012a), specifically the FRWIP (Morito 2008; Wheaton et al. 2007). Private insurance is not available in Saskatchewan in respect of flood for one's home, although provincial and federal government disaster assistance is (S4) and achieves its mandate of providing replacement costs to homeowners who suffer loss or damage as a result of flood.

5.7.3 Suasive Instruments

Suasive programmes appear to achieve their mandate. The FSP has facilitated practices improving water and soil quality, facilitating **adaptation** by increasing the **drought** resilience of producers (Wheaton et al. 2007). Many younger large agricultural producers are accessing information such as Drought Watch etc. posted on line (C11).

At the national level, Public Safety Canada coordinates the Hyogo Platform for all entities and organizations involved in emergency response. The effectiveness in this area has been reduced because of the driver of increased government austerity – the Canadian withdrawal from the provision of training through an emergency college, and increasing reliance on CSOs such as the Red Cross (C7, C8, C9).

5.7.4 Managerial Instruments

Interviewees regarded both the Boundary Water Treaty and the local watershed group source water planning as quite effective. To date there has not been a breach of the Treaty, and the management plans of the local water groups have been implemented and are being renewed. Many of these groups have expanded their source water protection to include anticipatory governance measures and Q3 climate change planning (S3). Interviewees also viewed the management instruments surrounding bankruptcy and transitioning agricultural producers to new liveihoods as very effective (C11).

5.8 Assessment of Learning

Turning to the question of how these instruments contribute to social learning, Table 5.3 shows that the most significant social learning in Saskatchewan relates to increasing climate variability (d&f) and the farm practices of agricultural producers (facilitated through the FWRIP and FSP). Examples include minimum tillage to sustain and conserve soil and soil moisture and conversion from cropping to permanent grass cover and community pastures (Diaz et al. 2009). Iterative participation between agricultural producers and government bureaucrats facilitated this learning (Hurlbert 2013). New technologies and practices (see Sect. 5.9.4) also indicate single loop learning and are partly a result of the driver of increasing trade liberalization and the effect of economic instruments. These changes have occurred not only at the individual producer level but also at the provincial or regional level when incorporated into FSP best practices.

In Saskatchewan no instrument facilitates the development or expansion of irrigation. As a result, this double loop learning (changing the assumption of dryland farming) is not as pervasive as in Alberta, and it only occurs in a few areas. This is partly due to the study area not having the same access to the SSRB (a relatively stable water source), the different culture of the Saskatchewan residents (preferring dryland farming), and the lack of institutional support (A4, A6). Without government assistance, even the most prosperous producers could not afford to maintain the entire dam and canal systems in addition to their own pivot systems (Diaz and

Table 5.3 D&F: Assessment of learning in Saskatchewan

Level/learning evidenced	Change in practice – single loop learning (one way information flow – Q1)	Change in assumptions or models – double loop learning (two way information flow – Q2 Q4)	Change in values, norms, world views or power dynamics – triple loop learning (iterative information flow – Q3)
Individual	Changes in agricultural practises, such as minimum tilling	–	–
Regional	FRWIP and FSP changing practises, e.g. reverting drylands to grasslands	–	–
National	–	–	–

Warren 2012a). The driver of government atrophy is compounding this as it withdraws from providing services to support current irrigation infrastructure (C1).

Instruments are lacking in Q3 and this absence correlates to a lack of triple and double loop learning. There are promising developments. Double loop learning is commencing with the advent of local watershed planning for source water protection in certain watersheds (see Sect. 5.9.2). These initiatives are broadly inclusive of all people willing to participate. Recent floods are changing the opinion of the public about climate change science (even in light of the driver of scepticism (S4)) and shows promise for more engagement with this Q3 issue.

5.9 Instrument Impacts on Livelihood Capitals

Instruments clearly impact society and the livelihood capitals available to producers to adapt to a changing climate (see Table 5.4). Each capital is discussed in turn.

5.9.1 Human Capital

D&f are harmful for mental health due to increased stress (Fletcher and Knuttila 2016). In times of flood the Red Cross was contracted to provide support services to flood victims including relocation, food, shelter, and counselling (S3). Agricultural producers exhibit a strong human capital in relation to education and adaptation strategies. The technological changes in farming practices responding to drought over the decades have occurred as a result of social connection. Not only do agricultural producers learn about adaptive measures at coffee row, but also at agricultural trade shows, during events sponsored by private companies (vendors and financers) (Hurlbert et al. 2014), and the group farming environmental plans described in Sect. 5.9.2.

5.9.2 Social Capital

The driver of changing farm size and urbanization is causing an exodus of young people and resulting aging of the local population (Davidson 2010) impacting the viability of education and health care services (Hurlbert et al. 2014), although currently strong producer networks exist (Diaz and Warren 2012a; C11). No instrument is addressing this change or retaining current social capital.

Currently 11% of Canada's irrigated land is situated in Saskatchewan (Sauchyn and Kulshreshtha 2010). Irrigation has been perceived as the primary adaptation of agriculture in the SSRB (S5). The diversified crop mix as a result of irrigation creates opportunities for economic development (Diaz and Warren 2012a) and builds a network of irrigators. An instrument encouraging this is missing.

Table 5.4 Assessment of main instruments

Type	Instrument	Effect	+ Impact on Capitals	(−) Impact on Capitals
Reg	Environmental and drainage liability (civil claim)	–		– Economic
	Drainage permits/fines	–		–Natural
	Emergency measures planning req	+	+ Social	
	GHG reduction	–		–Natural
But miss	Climate change mitigation and adaptation			
	Flood zone differentiation			
	Flood zone relocation			
Econ	FRWIP	+	+ Technology	=
	Loan instruments	+	+ Economic	
	Federal and provincial disaster assistance	+	+Economic	
	Growing Forward – Agri-programs	+	+ Economic	
	Agri-innovation	N.d.		
But miss	Irrigation expansion incentive			
	Private flood insurance			
	Water transfers			
	Carbon market			
Suas	FSP	+	+ Natural	
	Water security plan	+	+ Human	–
But miss	Participation instruments in international institutions UNCCD			
Mgt	Local watershed governance	+	+ Social	– Natural
	Irrigation association management	+	+ Social	
	Cooperation agreements for disaster recovery	+	+ Natural	
	Insolvency management	+	+ Economic	
	Boundary Water Treaty	+	+ Economic	
	Service contract Red Cross	+	+ Health	
But miss	Drought policy/drought research and planning program			
	Long term integrated water management plans			
	Inclusive, participatory development instruments			

Reg Regulatory, *Miss* Missing, *Econ* Economic, *Suas* Suasive, *Mgt* Managerial
++ Effective, + Moderately effective, – Ineffective, N.d. Not determined

Policies and instruments facilitating source water protection planning (carried out by local watershed groups) have increased social capital and also provide a local network for environmental farm planning through the Farm Stewardship Programme (FSP) (S3). Instead of individual producers completing an individualized environmental farm plan, a group of farmers in an area undertake a group environmental plan. This has allowed for more robust planning involving a bigger area, communal decision making, sharing of information, and has inhibited distrust of some producers in relation to the government's programme (Hurlbert 2013). This has assisted poorer farmers with less ability to access technological capital.

5.9.3 Economic Capital

Drought impacts agricultural income and also reduces income in the surrounding community (Kulshreshtha and Russell 1988).[5] Economic instruments (specifically the joint government-industry supported crop insurance program) support agricultural producer's retention of economic capital in times of drought (Wheaton et al. 2005). However, a significant number of producers interviewed in one study with low levels of capital (smaller producers), couldn't access these instruments for financial reasons (Hurlbert 2013).

Management (insolvency) instruments have assisted producers in retaining their home quarter and selling off or leasing their remaining land and finding off-farm income in mining, oil and gas, or agriculture related industry (see Diaz et al. 2009). Some have adapted by embarking on non-traditional initiatives including a bottled water business, a winery business, and a tourism enterprise (Diaz and Warren 2012b).

A significant cohort of agricultural producers are adapting through agricultural expansion taking advantage of such things as loan instruments, FRWIP, and Growing Forward instruments. This cohort doesn't describe their operation in the traditional sense of agricultural 'production' but as 'agri-business,' with focus on marketing and business practices including providing combining, production, and crop spraying services to other producers, producing ethanol, forming rail companies, etc. This group of producers has a greater level of knowledge about profitability, business practices, and diversification (C11).

5.9.4 Technological Capital

Economic and suasive instruments (FRWIP and FSP) have assisted producers adopt technological capital such as farm practices reducing soil tillage (often termed 'min till'), chemical fertilizers, herbicides, pesticides to allow continuous cropping (Warren 2016), shallow buried pipeline systems, and solar/wind powered dugout pumps to improve access to water for cattle (Hurlbert and Pittman 2014).

The large, capital-intensive agricultural producers have utilized credit instruments to acquire large expensive equipment (allowing self diagnostic, faster production) and drone technology to assist in fertilization and spraying (Melnychuk 2014). Suppliers continue to innovate, for example: changing tires to handle wetter soil without getting stuck (C11), enhancing seeds to increase yields, and micro-coating seeds with sulphur (fertilizer). Producers accessing the credit instruments also access software that tracks their yield and soil condition data via an application

---

[5] In Saskatchewan 17,803 jobs were lost in 2001 and 2002 as a result of the drought (Kulshreshtha et al. 2016).

on a smartphone that can be downloaded onto a computer to assist in future farm planning (C11). The new 2012 Agri-innovation instruments aim to capitalize on this technological innovation.

5.9.5 Natural Capital

There are both positive and negative impacts on natural capital. The driver of climate change may mean benefits for agriculture. Longer growing seasons, increasing the time period during the day when crops grow, and a warmer winter could all be positive for growth and yield of crops (Wheaton et al. 2010) and a shift to higher-value crops may result in increased gross margins (Easterling et al. 2007). New Agri-innovation instruments aim to capitalize on this.

Climate change, d&f are reducing natural capital through soil deterioration, wind erosion, and deterioration of grasslands (Wheaton et al. 2010). Floods cause damage to soils (higher salinity), worsened by larger farm equipment and the reduction of wetlands; historic roads with drainage ditches are being removed as farms expand (C11). The FRWIP assists producers with wells, buried pipelines and dug outs counteracting this driver (Hurlbert and Pittman 2014).

While irrigation in one northern area of the SSRB is thriving, irrigation agriculture in the study region of Rush Lake has been frustrated by three consecutive dry decades. Previously irrigators could rely on two full water allocations per year; now there is only water for one, sometimes only for half of the land irrigated, and sometimes for no one at all (Warren and Diaz 2012). Further south and west, irrigation developments have been completely abandoned due to lack of water (Warren 2013).

Missing instruments are listed in Table 5.4 and include one addressing climate change and inclusive participatory development.

5.10 Re-designing Instruments with the ACW

This chapter explored how Saskatchewan's institutional policy framework impacts capitals of agricultural producers given local and national drivers. Utilizing these findings and the ACW, the adaptive capacity of Saskatchewan's institutional framework is assessed. Based on this assessment, recommendations are made for policy redesign to improve the adaptive capacity of the governance institutions of climate change, d&f itemized in Table 5.5.

Saskatchewan's governance system in respect of agricultural producers, climate change and response to d&f is depicted in the ACW in Fig. 5.3. Starting from the top and moving clockwise, room for improvement exists in the aspect of **variety**. The construction of risk should be expanded to recognize the interrelated policy problems of climate change, adaptation, d&f from a variety of problem frames, not

Table 5.5 Redesign options for Saskatchewan

Analysis unit	Description
Ineffective instruments	GHG reduction
	Drainage permits, liability, fines, flood zone mapping, building codes
	Hyogo platform
Missing instruments	Climate change mitigation and adaptation, carbon market
	Participation in UNFCCC, UNCCD
	Flood zone differentiation, flood zone relocation, private insurance
	Irrigation expansion incentive
	Inclusive, participatory development instruments
	Water transfer, drought policy, drought research and planning program
	Integrated water resource management
Unaddressed drivers	Government austerity
	Climate change, climate scepticism
	Aging infrastructure
	Growing inequality
	Deteriorating ecosystems
Weak adaptive capacity dimension	Equity
	Variety (problem frames and solution, redundancy)
	Learning capacity (discuss doubts)
Constrained livelihood capital	Social capital – urbanization, exodus of young producers, larger production units
	Natural capital – climate change deteriorating ecosystem services

just in relation to reactive measures to d&f. Very little redundancy exists in Saskatchewan's instrument suite; there is some diversity of actors across multi-levels of government.

Producers in Saskatchewan have adapted to climate all of their lives, it is part of their habit; but this **learning capacity** is predominantly single loop learning. Irrigation (double loop learning) is historic; instruments supporting its development are no longer in existence. Building new irrigation infrastructure is increasingly costly, and unattainable without government economic instruments. Climate change increases the risk associated with these investments as projections of less available water in the rivers potentially strand infrastructure investment.

Economic instruments have supported a strong agricultural economy (albeit one being further differentiated between small and big producers). These economic instruments as well as strong social capital in Saskatchewan have encouraged significant technological learning and leave **room for autonomous change**. The management instruments surrounding bankruptcy as well as strong human capital in terms of education allow changes in farming, alternative career choices, and agribusiness in the study area and provide the ability to improve. The regulatory instruments surrounding water and its management have also provided flexibility in responding to drought (S2, S3).

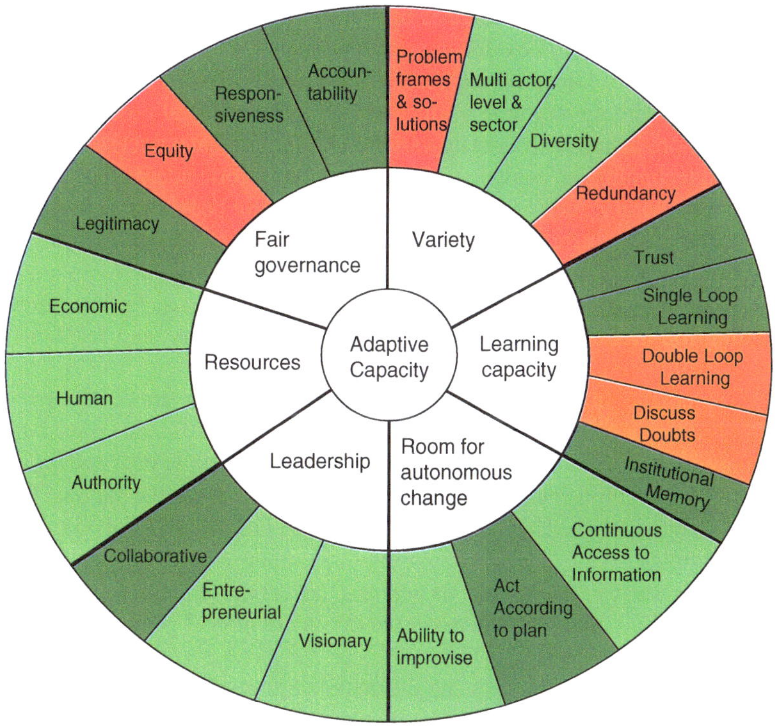

Green	Light Green	Yellow	Light Orange	Red
Very High	High	Medium	Low	Very Low

Fig. 5.3 Saskatchewan's ACW

Local watershed groups and provincial social capital leverage **leadership** collaboratively by building water resilience and improving environmental practices through the FSP. Producers are visionary, strong entrepreneurial **leaders**. The suite of Agri-Innovation, credit availability, and bankruptcy/debt management also allow producers to change in an autonomous, entrepreneurial manner. New instruments (Agri-innovation) are aimed at facilitating producer innovation and improvisation, but it is too soon to assess their effectiveness.

In the industrialized country of Canada, provincial governments do have **resources.** There is strain on resources due to the driver of government atrophy. One weakness exists in that municipalities have the least resources and yet are tasked with response to d&f; this reduces redundancy of instruments and tools. Inequity in community resources and access to knowledge and information about drinking water exists. Local municipalities receive assistance funds late, have few resources, few staff, and their population and tax base are declining (S4). Local bridges and roads require rebuilding at municipal expense after floods and smaller municipalities have more difficulty.

The retreat of federal government has been met by a surge of CSOs and NGOs, but resulted in fewer services for marginalized and Aboriginal people (S1). Responsiveness, part of **fair governance,** is taxed. **Equity** is also a growing concern for municipalities and agricultural producers. Bigger agricultural production units are generally more adaptive to climate variability having greater access to human, economic and technological capital and leveraging the historic adaptations of farmers over the last century. As a result, bigger farm units have been a trend for the last several decades and an increasing spread between large farms and small farms, and their incomes, is apparent.

Redesign options for Saskatchewan include adopting the missing instruments, strengthening the ineffective instruments, addressing all drivers, and focusing on constrained livelihood capitals identified in Table 5.5.

Most problematic in Saskatchewan's suite of instruments are: drainage and flooding instruments (which are unutilized), building standards and codes which don't account for climate change, and the lack of coordinated and integrated drought research and planning programmes at the regional (provincial) or the national level. This is significant given the large economic, environmental and social damages that result from drought.

Missing instruments are needed to address the driver of climate scepticism, involve participation of Saskatchewan people in environmental governance, and to implement the UNFCCC and UNCCD.

The deficit of instruments addressing the driver of government atrophy and concomitant reduction of the federal government mandate and presence has had serious impacts. There is no longer supported emergency management training. There are no programmes assisting with irrigation expansion (considered by many a form of adaptation) (S3). There is no federal long-term water management plans. As a result there is very little understanding of changing groundwater levels (Wittrock et al. 2010) or understanding of the nexus of agriculture and water (Wittrock and Kulshreshtha 2011) and energy. Most concerning, are the lack of instruments addressing the driver of future climate change, climate change scepticism, and deteriorating ecosystem services, because these policy problems are not cognized. The increasingly absent federal government signals a loss of adaptiveness in the institutional governance of d&f in Saskatchewan.

Governance in Saskatchewan is adaptive in the sense that agricultural practices and institutions have changed and adapted to historic d&f maintaining a vibrant agricultural sector. However the interconnections of people, livelihood capitals and instruments are not always considered in a holistic, inclusive manner (for all people) and long term and transitionary time frames given climate change are lacking. Addressing the identified gaps in the institutional governance framework, especially those of climate change and its embedded problems, could improve adaptive governance and build capacity of rural agricultural producers and their communities into the future.

References

AAFC (2002) Drought risk management plan for Saskatchewan, 2002. In: Canada-Saskatchewan MOU on water committee minutes, June 24, 2002. Agriculture and Agri-Food Canada–PFRA, Regina

AAFC (Agriculture and Agri-Food Canada) Drought Watch (2015) Drought watch. Available at: www.agr.bc.ca/eng/?id=1326402878459. Accessed 31 Mar 2015

Barrow E (2010) Climate change scenarios for the Prairie Provinces. In: Sauchyn D, Diaz H, Kulshreshtha S (eds) The new normal. Regina, Canadian Plains Research Center

Bruneau J, Corkal DR, Pietroniro E, Toth B, Van der Kamp G (2009) Hu-man activities and water use in the South Saskatchewan river basin. Prairie Forum 34(1):129–152

Byrne J, Kienzle S, Sauchyn D (2010) Prairies water and climate change. In: Sauchyn D, Diaz H, Kulshreshtha S (eds) The new normal. Regina, Canadian Plains Research Center

Canada (2002) Combatting desertification: building bridges. CIDA, Ottawa. Available at: ttps://www.google.ca/webhp?sourceid=chrome-instant&rlz=1C1ASUT_enCA545CA561&ion=1&espv=2&ie=UTF-8#q=Canada+submission+to+un+desertification+2002. Accessed 30 Apr 2015

Canada (2015) Intended nationally determined contribution. Available at: unfccc.ca. Accessed 12 Apr 2015

CAUT (2013) A spotlight on the many ills that plague scientific pursuit in Canada. CAUT Bulletin 60(10) December 2013. p 1. Ottawa

CBC (2011) Canada pulls out of Kyoto Protocol. Posted December 2011 www.cbc.ca. Accessed 12 Apr 2015

CEPA (1999) Reduction of carbon dioxide emissions from coal-fired generation of electricity regulations, P.C. 2012–1060 August 30, 2012 pursuant to the Canadian Environmental Protection Act, 1999, S.C. 2004, c. 15, s. 31

Davidson D (2010) The social landscape. In: Sauchyn D, Diaz H, Kulshreshtha S (eds) The new normal. Regina, Canadian Plains Research Center

Diaz H, Warren J (2012a) RCAD – Rural Communities Adaptation to Drought, Research Report, August 2012. University of Regina and CPRC Press, Regina

Diaz H, Warren J (2012b) Defying Palliser. CPRC Press, Regina

Diaz H, Haradats M, Barret-Diebert P (2009) IACC Institutional Adaptations to Climate Change Project, final report December 2009, comparative study of dryland river basins in Canada and Chile. CPRC Press, Regina

East V, Pittman J, McRae K, Soulodre E (2012). Drought and excessive moisture pre-paredness plan. Upper Souris River Watershed. Available at: http://www.parc.ca/rac/fileManagement/upload/Upper_Souris_DEM_plan_July%209_2012.pdf. Accessed 12 Apr 2015

Easterling WE, Aggarwal PK, Batima P, Brander KM, Erda L, Howden SM, Kirilenko A, Morton J, Soussana JF, Schmidhuber J, Tubiello FN (2007) Food, fibre and forest products. In: Parry ML, Canziani OF, Palutikof JP, van der Linden PJ, Hanson CE (eds) Climate change 2007: impacts, adaptation and vulnerability. Contribution of working group II to the fourth assessment report of the intergovernmental panel on climate change. Cambridge University Press, Cambridge, pp 273–313

EC (2012) Backgrounder. Reduction of carbon dioxide emissions from coal-fired generation of electricity regulation. Accessed 2 Jan 2013. http://www.ec.gc.ca/default.asp?lang=En&n=5C4438BC-1&news=D375183E-0016-4145-A20B-272BDB94580A. Accessed 12 Apr 2015

Ekos (2013) So what's really bothering you Canada: (and what do you want your federal government to do about it?) Available at: Ipolitics.ca Informative, insightful, influential. Ottawa

Ferguson (2013) Chapter 8: Disaster relief for producers – agriculture and agri-food Canada Office of the Auditor General of Canada, Ottawa. Available at: http://www.oag-bvg.gc.ca/internet/English/parl_oag_201311_08_e_38802.html

Fletcher AJ, Knuttila E (2016) Chapter 7: Gendering change: Canadian farm women respond to drought. In: Diaz H, Warren J, Hurlbert M (eds) Vulnerability and adaptation to drought: the Canadian Prairies and South America, ISSN 1919–7144. calgary. University of Calgary Press, Alberta, pp 159–178

FPTGC (Federal, Provincial and Territorial Governments of Canada) (2010). Canadian biodiversity: ecosystem status and trends 2010. Canadian Councils of Resource Ministers, Ottawa. vi + p 142

GC (2012) Backgrounder on agriinnovation, agrimarketing and agricompetitiveness programs. Government of Canada, Ottawa

GC (2014) Canada's sixth national report on climate change, 2014. Queen's Printer, Ottawa. Available at: http://unfccc.int/files/national_reports/annex_i_natcom/submitted_natcom/application/pdf/nc6_can_resubmission_english.pdf. Accessed 15 Sept 2014

GC (Government of Canada) (2013a) Disaster Financial Assistance Arrangements (DFAA), effective date January 1, 2008. Available at: www.publicsafety.gc.ca/cnt/mrgnc-mngmnt/rcvr-dsstr/dsstr-fnncl-ssstnc-rrngments/index-eng.aspx. Accessed 22 Apr 2015

GC (Government of Canada) (2013b) Canada's national disaster mitigation strategy, updated August 19, 2013, Available at: www.publicsafety.gc.ca/cnt/rsrcs/pblctns/mtgtn-strtgy/index-eng.aspx. Accessed 14 Apr 2015

GC (Government of Canada) (2013c) Canada's platform for disaster risk reduction. n.d. Available at: www.pulicsafety.gc.ca/cnt/mrgnc-mngmnt/dsstr-prvntn-mtgtn/pltfrm-dsstr-rsk-rdctn/index-eng.aspx. Accessed 14 Apr 2015

Graveland B (2013) Brazilian Company Buys XL Foods Plant involved in E. Coli Meat Recall. Macleans, January 9, 2013

GS (2011) Farm and ranch water infrastructure. Available at: http://www.agriculture.gov.sk.ca/FRWIP_2009. Accessed 29 Sept 2011

GS (2012) About government relations. Available at: http://www.gr.gov.sk.ca/about/Accessed 12 Apr 2015

GS (2013a) Provincial disaster assistance program, Eligible Expenses. Available at www.gr.gov.sk.ca/PDAP. Accessed 12 Apr 2015

GS (Government of Saskatchewan) (2013b2013a) Growing forward 2, Agirculture Demonstration of Practices and Technolgoies (ADOPT) available at: www.agriculture.gov.sk.ca/GF2-ADOPT. Accessed 12 Apr 2015

GS (2015) Individual and group plans are provided for. Environ Farm Plan. Available at: http://www.agriculture.gov.sk.ca/GF2-EnvironmentalFarmPlan. Accessed 31 Mar 2015.

Hamadijan D (2013) Canada has become the first country to drop out of the UN convention to combat desertification. CTV National News: Canada on the world stage. Available at: http://www.ctvnews.ca/politics/canada-only-un-member-to-pull-out-of-droughts-and-deserts--convention-1.1214065. Accessed July 2014

Henstra D (2011) The dynamics of policy change: a longitudinal analysis of emergency management in Ontario, 1950–2010. J Policy History 23(3):399–428

Hill A (2014) Science abandoned, scientists muzzled. The Leader Post. Friday January 10, 2014, p. A5

Hurlbert M (2009) Comparative water governance in the four Western Provinces. Prairie Forum, a special edition on climate change 34(1):45–77

Hurlbert M (2013) Learning, participation, and adaptation: exploring agri-environmental programmes. J Environ Plann Manag. Published online: http://dx.doi.org/10.1080/09640568.2013.847823. Accessed 20 Jan 2015

Hurlbert M, Pittman J (2014) Exploring adaptive management in farm programs in Saskatchewan. J Nat Resour Policy Res. doi:10.1080/19390459.2014.915131

Hurlbert M, Diaz H, Knutilla E, Hernani B (2014) Community vulnerability. Presented at the Canadian Association of Geographers in Brock University, St., Catherines, Ontario, Canada, May 28, 2014

Hurlbert M, Testfamariam Y, Andrews E (2015) Governing water, deliberative institutions and adaptation. Research Report

IPCC (2014) Summary for policymakers. In: Edenhofer O, Pichs-Madruga R, Sokona Y, Farahani E, Kadner S, Seyboth K, Adler A, Baum I, Brunner S, Eickemeier P, Kriemann B, Savolainen J, Schlömer S, von Stechow C, Zwickel T, Minx JC (eds) Climate Change 2014, Mitigation of Climate Change. Contribution of Working Group III to the Fifth Assessment Report of the Intergovernmental Panel on Climate Change. Cambridge University Press, Cambridge/New York

Johnstone B (2012) Farms getting bigger, but farmers fewer and older, the Leader-Post, May 11, 2012, p C1

Kulshreshtha SN, Russell KD (1988) Economic impact assessment of irrigation development and related activities in Manitoba. Canad Water Resour J 28(1):53–67

Kulshreshtha S, Wheaton E, Wittrock V (2016) Chapter 4: The impact of the 2001–2002 drought in rural Alberta and Saskatchewan, Canada. In: Diaz H, Warren J, Hurlbert M (eds) Vulnerability and adaptation to drought: the Canadian Prairies and South America, ISSN 1919–7144. University of Calgary Press, Calgary, pp 79–106

Melnychuk M (2014) Drone technology aids Sask. Farmers. Section D, D1, The Leader Post, Wednesday 24 Sept 2014

Morito B (2008) Institutional values and adaptive capacity to climate change, 2008. Institutional Adaptation to Climate Change Research Project, University of Regina. http://www.parc.ca/mcri/pdfs/papers/iacc085.pdf. Accessed 12 Apr 2015

Office of the Auditor General of Canada (1994) Audit of farm income stabilization programs. The Queen's Printer, Ottawa

PSC (Public Safety Canada) (2014). Disaster Financial Assistance Arrangements (DFAA). Available at: www.publicsafety.gc.ca/cnt/mrgnc-mngmnt/rcvr-dsstrs/dsstr-fnnd-ssstnc-rrngmnts/index-eng.aspx. Accessed 12 Apr 2015

Remax (2014) Farm Report 2014. Available at: http://download.remax.ca/PR/FarmReport/FinalReport.pdf. Accessed 16 Apr 2015

Rowan K, Pittman J, Wittrock V, Unvoas A (2011) Drought and excessive moisture preparedness plan, March 2011. Canada, Ottawa. Available at: http://www.parc.ca/rac/fileManagement/upload/SCCDEM%20Plan%20March%2029%20Final.pdf. Accessed 12 Apr 2015

Sandink D, Kovacs P, Oulahen G, McGillivray G (2010) Making flood insurance available for Canadian homeowners. A discussion paper, November, 2010. Institute for Catastrphic Loss Reduction & Swiss Reinsurance Companies Ltd, Toronto

Sauchyn D, Kulshreshtha S (2010) Chapter 17: agricultural adaptation through irrigation. In: Diaz H, Warren J (eds) Vulnerability and adaptation to drought in the Canadian Prairies. University of Calgary Press, Calgary

Statistics Canada (2011). Farm debt outstanding: agriculture economic statistics. Minister of Industry, Ottawa, 21–014-X, vol 10, no.2 available: http://www.statcan/gc/ca/pub/21-014-x/21-014-x2011002-eng.pdf. Accessed 14 Sept 2014

Statistics Canada (2012) Snapshot of Canadian agriculture: farm and farm operator data Ottawa: statistics Canada, 95–640-XWE. available: http://www.statcan.gc.ca/pub/95-640-x/2012002/01. Accessed 14 Sept 2014

SWSA (Saskatchewan Water Security Agency) (2012) 25 Year Saskatchewan water secuirty plan. Queen's Printer, Regina. Available at: http://www.gov.sk.ca/adx/aspx/adxGetMedia.aspx?mediaId=1798&PN=Shared. Accessed 17 Sept 2014

Toth B, Corkal DR, Sauchyn D, van der Kamp G, Pietroniro E (2009) The natural characteristics of the South Saskatchewan River Basin. In: Marchildon G (ed) A dry Oasis: institutional adaptation to climate on the Canadian plains. CPRC Press, Regina, pp 95–127

UNFCCC (2013) FCCC/CP/2013/10/Add.1 Available at: http://unfccc.int/documentation/documents/advanced_search/items/6911.php?priref=600007788. Accessed 15 Sept 2014

Wandel J, Pittman J, Prado S (2010) Rural vulnerability to climate change in the south Saskatchewan river basin. In: Sauchyn D, Diaz H, Kulshreshtha S (eds) The new normal. The Canadian Prairies in a changing climate. CPRC Press, Regina, pp 245–258

Warren J (2013) Rural water governance in the Saskatchewan Portion of the Palliser Triangle: an assessment of the applicability of the predominant paradigms. In: A Thesis Submitted to the Faculty of Graduate Studies and Research In partial Fulfillment of the Requirements for the Degree of Doctor of Philosophy, 2013. University of Regina, Regina

Warren J (2016) Chapter 5: The min till revolution and the culture of innovation. In: Diaz H, Warren J, Hurlbert M (eds) Vulnerability and adaptation to drought: the Canadian Prairies and South America, ISSN 1919–7144. University of Calgary Press, Calgary, pp 107–132

Warren J, Diaz H (2012) Defying Palliser: stories of resilience from the driest region of the Canadian Prairies. CPRC Press, Regina

Wheaton E, Wittrock V, Kulshreshtha S, Koshida G, Grant C, Chipanski A, Bonsal B (2005) Lessons learned from the Canadian drought years of 2001 and 2002: Synthesis Report. SRC Publication No. 11602-46E03. Saskatchewan Research Council, Saskatoon

Wheaton E, Koshida G, Bonsal B, Johnston T, Richards W, Wittrock V (2007) Agricultural adaptation to drought (ADA) Canada: the case of 2001 and 2002. Saskatchewan Research Council, Publication No. 11927-1E07, Regina

Wheaton E, Kulshreshtha S, Wittrock V (2010) Chapter 16: Assessment of the 2001and 2002 drought impacts in the Prairie Provinces, Canada. In: Warren DH, J. (eds) Vulnerability and adaptation to drought in the Canadian Prairies. University of Calgary Press, Calgary

Wheaton E, Wittrock V, Koshida G, Siemens E, Smeh D (2011) What are the benefits of drought? Concepts and examples to promote effective climate adaptation information services, prepared for adaptation and impacts research section, environment Canada, SRC Publication No. 13077-1E11, March 2011. Saskatchewan Research Council, Saskatoon

Wheaton E, Sauchyn D, Bonsal B (2016) Chapter 3: Future possible droughts. In: Diaz H, Warren J, Hurlbert M (eds) Vulnerability and adaptation to drought: the Canadian Prairies and South America, ISSN 1919–7144. University of Calgary Press, Calgary, pp 59–76

Wittrock V, Kulshreshtha S (2011) Social and economic indicators of vulnerability: preliminary community profiles. Main Report with Summary Tables

Wittrock V, Wheaton E, Siemens E (2010) More than a close call: a preliminary assessment of the characteristics, impacts of and adaptations to the drought of 2008–2010 in the Canadian Prairies, prepared for environment Canada adaptation and impacts research division (AIRD), SRC Publication No. 12803-1E10, March 2010. Saskatchewan Research Council, Saskatoon

Wyld A (2014) Canada “frank” on climate change: PM. The metro. The Associated Press, Tuesday June 10, 2014, p 03

Yalnizyan A (2013) Study of income inequality in canada – what can be done, Presentation to the House of Commons Standing committee on Finance. Available at: https://www.policyalternatives.ca/publications/reports/study-income-inequality-canada%E2%80%94what-can-be-done#sthash.Q1YYvuNR.dpuf. Accessed 14 Apr 2015

Laws

(EPA) The Canadian Environmental Protection Act. RSC 1985. C. 16 (4th Supp)

Passenger Automobiles and Light Trucks, Heavy-Duty Vehicles, Renewable Fuels and Reduction of CO2 Regulations pursuant to the Canadian Environmental Protection Act, 1999, S.C. 2004. c. 15, s. 31

Reduction of Carbon Dioxide Emissions from Coal-fired Generation of Electricity Regulations, P.C. 2012–1060 August 30, 2012 pursuant to the Canadian Environmental Protection Act, 1999, S.C. 2004. c. 15, s. 31

The Bankruptcy Act. RSC 1985. C. B-3
The Boundary Water Treaty of 1909
The Canadian Environmental Assessment Act. SC.l 1992. C. C-11
The Constitution Act. 1982, being Schedule B to the Canada Act, 1982, 1982, C 11
The Emergency Planning Act. SS 1989–90, c. E-8.1
The International Boundary Waters Treaty Act. RSC 1985. C. I-17
The Management and Reduction of Greenhouse Gases Act, not yet proclaimed
The Municipalities Act. SS 2005. C. M-36.1
The Saskatchewan Crop Insurance Corporation Act. SS 2012. C. S 12.1
The Water Security Agency Act. SS 2005. C. W-8.1

Chapter 6
Case Study Alberta, Canada

6.1 Introduction

This chapter explores the impact of Alberta's institutional policy framework on agricultural producers and is structured similarly to Chap. 5. The effectiveness of the main instruments responding to d&f are evaluated and analysed in relation to effectiveness, policy framing, impact on livelihood capitals, and dimensions of the ACW, in the context of drivers.

In Alberta adaptive governance shows promise. However significant disconnections between the issues of climate change, d&f and inability to achieve a modicum of IWRM inhibit many positive adaptive governance developments. This chapter shows that Alberta has the greatest variety of instruments and many organizations relating to water, climate change, d&f. However, the current instruments addressing climate change are ineffective. The lack of adaptive management and anticipatory governance methods together with trade liberalization and growing energy demand (Alberta makes a significant contribution to the oil and gas industry) jeopardizes Alberta's natural capital as GHG emissions rise and ecosystem services reduce. Climate denial perhaps contributes to this. Clearly adaptive management and anticipatory governance are important components of adaptive governance as their absence in Alberta is associated with an ineffective, but very populated suite of d&f instruments.

Significant learning has occurred in Alberta's past with a special administrative board created decades ago in a particularly drought prone area to manage land, the creation of irrigation, and implementation of a limited water transfer arrangement responding to drought in 2001. Current instruments will not protect agricultural livelihoods from forecasted climate change that includes longer more significant droughts and it isn't clear that social learning will continue in the future. Zero loop learning has occurred historically in relation to the construction of dams and in relation to flood.

M.A. Hurlbert, *Adaptive Governance of Disaster*, Water Governance - Concepts, Methods, and Practice, DOI 10.1007/978-3-319-57801-9_6

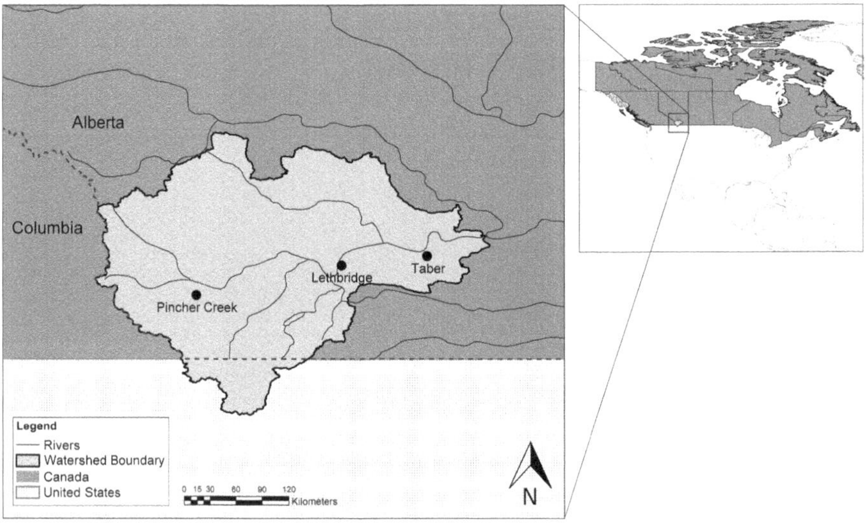

Fig. 6.1 Study regions in Alberta, Canada

The High River flood of 2013 illuminated the ineffectiveness of regulatory flood instruments, the absence of flood insurance, and lack of comprehensive flood management. Poor people experienced the most severe flood impacts. Marginalized people have been excluded from development decisions (the zero loop learning of the Oldman dam described in Sect. 6.8 exemplifies this). Although Alberta has many participation instruments, they have been ineffective in tackling the drivers of climate change and growing inequality.

6.2 The Case Study Area

The Alberta study region is located within the same basin as the Saskatchewan study area, the SSRB, but further west and closer to the Rocky Mountains. The climate is continental, sub-human to semiarid, with short hot summers, cold long winters, low levels of precipitation, and high summer evapotranspiration (Sauchyn and Kulsheshtha 2010).

Southern Alberta is a prosperous, developed, vibrant area consisting of a diversity of ecosystems (Byrne et al. 2010). Figure 6.1 illustrates the study communities in Alberta of Taber and Pincher Creek. Development of irrigation as far back as 1915 as well as a significant oil and gas industry set the stage for a prosperous community (Bruneau et al. 2009). A range of diversified specialty crops are grown and a significant industry of local food processing exist for these crops. Dryland areas are utilized for not only seeded crops but 56% of farms are livestock operations (*ibid.*). Whereas Saskatchewan has considerable experience surrounding drought, Alberta has significant recent experience surrounding flood (with the High River Flood in 2013).

6.3 Main Drivers Impacting Rural Agricultural Producers

There are three main themes in relation to drivers: climate change, global–local economics, and the relationship between the two (see Table 6.1).

Climate change in Alberta is resulting in more variable weather with increasing length and intensity of drought and increased rainfall. Climate change impacts involve drier conditions, more droughts impacting quantity and quality of water, and increased climate uncertainty (IPCC 2014). Climate change models provide that annual mean temperature will increase between 0 °C and 2 °C in the 2020s and

Table 6.1 Drivers impacting agricultural producers' capitals

	Driver – direct		Driver – indirect	
Category	Local	National	Local	National
Demographic	Increasing size of farms	–	Urbanization/population growth	–
	Shortage of farm labour (Wandel et al. 2010)		Aging producers	
Economic	Escalating land prices (Remax 2014)	–	Strong oil and gas economy	Increasing exposure to global markets
	Changing food prices			World recession 2008
	Changing net private capital flows			Tempered economic development
				Reduction of size and budget of government
Political Economy	Increasing cost of inputs	Reduction of government services and programmes	Aging infrastructure (roads, dams)	Growing inequality (Yalnizyan 2013)
				Increasing trade liberalization
				Increasing demand for energy
Social	Climate change denial	–	–	Priority of economy over environment
Natural	Climate change	–	–	–
	Deteriorating ecosystems (FPTCG 2010)			

between 2 and 4 °C in the 2080s (Barrow 2010). There will be a slight decrease in minimum annual precipitation (between 0 and 10%) across the prairie region in all time periods with the most severe impacts experienced in the study region (*ibid.*). This will be combined with longer, warmer winters with reduced snowpack accumulations and glaciers in the mountains changing stream flow (Byrne et al. 2010). Some areas such as the Battle River in the SSRB have already experienced a collapse of stream flow resulting in severe water restrictions and shutdown of power plants (*ibid.*).

Alberta shares the **global and local economic drivers** of Saskatchewan (see Sect. 5.3). Reduction in government programmes together with growing trade liberalization is increasing uncertainty. The federal government has disbanded the single desk wheat selling entity and a transportation subsidy on rail shipments (Harvey 2005). Another example is the disbandment of the federal Emergency Preparedness College and withdrawal of federal funding for assisting local government people in training, establishing contacts, and learning how to plan for emergencies (A1).

Large multinational corporations increasingly dominate multiple links in the food chain.[1] Alberta's farms are getting larger,[2] run by older people,[3] and more profitable. The growing gap between poor and wealthy (Yalnizyan 2013) exists in the agricultural sector. Although the total number of farms is decreasing, big farms are getting bigger and making more money, while smaller farms make less.[4] A higher debt load (Statistics Canada 2011) results in a tighter squeeze between high input costs and low commodity prices.

As in Saskatchewan, the **relationship between climate change and economics** is strained. The former Prime Minister's comments are equally applicable to Alberta (see Sect. 5.3). Alberta has a unique place within this context of climate denial. The provincial government representative of High River campaigned on climate change denial and continued in his denial even after the High River flood (Rocher 2013). Federal funding cuts to environmental science (CAUT 2013) impact Alberta oils sands in a unique way. One specific example is the loss of eleven regional libraries of aquatic research and resulting disposal of an environmental study in relation to building an oil pipeline in the 1960s and 1970s (Galloway 2014). This research would have acted as a baseline for current proposed pipelines in the same area. The natural environment is deteriorating and there is a loss of ecosystem services (FPTGC 2010).

---

[1] e.g. 95% of beef processing in Canada is done by two major corporations: Cargil and JBS Food Canada, a subsidiary of a Brazilian company (Graveland 2013).

[2] Further, five % of farms generate nearly half of total farm cash receipts (Johnstone 2012). The number of farms have declined by 10.3% nationally and 16.6% in Saskatchewan between 2006 and 2011 (a trend since 1941) (Statistics Canada 2012; Johnstone 2012).

[3] Nearly half of farmers are 55 years of age or older a (Statistics Canada 2012; Johnstone 2012).

[4] There was a 12.5% decrease in the number of farms since 2006 and the number of farms with over one half million dollars in gross farm receipts in 2010 increased by 18%; those with less decreased by 15.1%. Farms over this mark represented 10.35% of all farms in the province in 2010, but accounted for 70.7% of total provincial gross farm receipts (Statistics Canada 2012).

6.4 Institutions (Organizations) That Build Capacity for Climate Change, d&f

The main institutions (organizations) are identified here in turn. As in Saskatchewan, the most important government institutions are provincial and even the federal institutions provide programming via provincial counterparts. Local institutions, however, are most important for immediate emergency assistance.

Alberta Environment and Sustainable Resource Development (AESRD) has several relevant functions: overseeing the **Climate Change** Adaptation Framework in conjunction with Alberta Agriculture and Rural Development (AARD); managing water and networking with the Alberta Water Council and other entities with mandates relating to water; overseeing development (including oil-sand); enforcing environmental laws and regulations; and overseeing land use management plans (AESRD 2015).

AARD together with Ag Canada provide programming for agriculture. The Alberta Drought Management Committee coordinates response to **drought**. The responsibility of planning for and responding to an extreme event such as a **flood** resides with the municipality or local government. When the local government is overwhelmed (as in the case of the High River flood in 2013), the Alberta Emergency Management Agency (AEMA) becomes involved. The AEMA sets up a communication centre for coordination with an NGO Council and agencies such as the Royal Canadian Mounted Police (RCMP) (A2). Funding for disaster recovery is provided by Public Safety Canada and paid through the provincial AEMA (A2).

More CSOs and NGOs exist in Alberta than Saskatchewan. As in Saskatchewan, CSOs have mandates in relation to a particular subject (water, the environment, safety etc.) and don't integrate comprehensively climate change, d&f (except for PARC).

Private organizations include not only large ranches, but also intensive livestock operations, and food processing companies in the study area. Alberta has more producers using irrigation (and associated irrigation districts) as historically there were more instruments developing irrigation (A4). Insurance companies are also a key component of the private organizational institutional structure.

Alberta participates in the same interconnecting institutions as Saskatchewan (see Sect. 5.4); as a result Alberta must share its water with downstream provinces and retains no power from its geographic positioning at the headwaters (PPWB 2003). Other coordinating organizations include irrigation districts that obtain a licence and then internally govern sub-licenses within their district (A3).

6.5 Instruments Responding to Climate Change, d&f

The same instruments of the Canadian government listed in Saskatchewan exist in Alberta, as well as a few common provincial instruments (such as regulatory water licensing). Several additional relevant provincial instruments also exist (see

Table 6.2 Institutional water instruments

Principle	Description
Principle under which water is managed	Most beneficial use
Allocation of water rights	Statutorily legislated model with some water trading
Priorities	First in time, first in right principles
Water Market	Transfers of water independent of land allowed in study area of SSRB
Water allocation dispute resolution	AESRD Minister responsible, then court litigation
Potable water accountability	Local providers via legislated drinking water quality standards, wastewater treatment controls, and reporting requirements
Governance Accountability	Environmental Appeal Board hears drinking water disputes
Water price	Tariff regulated by Alberta Utilities Commission

Source: Water Act 2000; Hurlbert 2009b

Table 6.4 which compares the main instruments in Sect. 3.3.3). Alberta has relevant regulatory, economic, suasive, and managerial instruments, but iterative climate change planning and programs of adaptive management are missing.

6.5.1 Regulatory Instruments

Alberta has many regulatory measures pertaining to water governance, climate change, d&f. Alberta's Water Act legislates the priority system relating to surface and ground water as a first in time, first in right scheme structured within Q1. The legislation provides both a regulated framework and also the ability to transfer water for compensation, or a market interest. Table 6.2 outlines water instruments in Alberta.

In Alberta **climate change** legislation has existed since the Climate Change and Emissions Management Act (2003), a precursor to Alberta's Climate Change Strategy (2008) occupying Q3, Q1, and Q4. This Act established a carbon offset market, consumer energy efficiency rebates, and GHG reporting and reduction programmes. Large emitters (making up 70% of Alberta's emissions) are required to reduce their emissions by 12% from 2003 emission levels. Alberta's climate change strategy focuses on both CCS and 'greening' energy production. It is aiming to reduce GHG emissions by 14% below 2005 levels. Canada limits GHG emissions by coal fired power generation (EC 2012).

Municipalities shoulder the majority of obligations in respect of **flooding**. Alberta's Emergency Management Act R.S.A., c. E-6.8 provides that all municipalities shall have emergency response plans in place. Infrastructure (such as roads and bridges) must be designed for a one in one hundred years (1/100) event (A2, A6); flood zone building restrictions exist (but not in the case study area).

6.5.2 Economic Instruments

Alberta leads in the development of a water market in Canada. The Alberta market model was developed and touted as advancing the goals of efficient allocation of water interests and conservation by encouraging the transfer of surplus interests (Percy 2004). The market only applies within a limited area, on a limited basis (Hurlbert 2009b, c). Water licences, for the most part, continue to be regulatory instruments, granted to property owners in respect of a parcel of land and then transferred with the land, all managed in a Q1 manner.

Currently Alberta has no federally funded initiatives for **climate adaptation**. The carbon market created by regulation (see Sect. 6.5.1) is arguably an economic instrument addressing climate mitigation. In respect of adaptation, the same federal government 'Growing Forward 2' instruments (see Sect. 5.5.2) available to Saskatchewan agricultural producers are available in Alberta. These programmes (AgriInvest, AgriStability, AgriInsurance, AgriRecovery) provide a range of technocratic Q1 income stability measures in the event of d&f, but are not framed in relation to climate change or adaptation. The new programmes of AgriInnovation, AgriMarketing, and AgriCompetitiveness as well as two additional provincial programmes: irrigation efficiency (AARD 2015a) and loans for disaster assistance (AG 2010), address the driver of trade liberalization.

Alberta and Saskatchewan have similar disaster assistance programmes: Insurance companies provide business insurance; flood insurance is not available for flood and only in some places for sewage backup. Agricultural producers rely on government for some insurance and loans during disaster. Disaster assistance is cost shared between the provincial and federal governments and provides residents and municipalities up to 90% of expenses.

6.5.3 Suasive Instruments

Federal government instruments of information (Drought Watch), and environmental farm planning (the Farm Stewardship Programme (FSP)) are available to agricultural producers. The Hyogo platform is accessed by government and CSOs for information sharing and preliminary steps are underway of establishing a federal disaster strategy.

Alberta also provides agro-climatic information (AARD 2015b) and has centralized this in one web site. This web site provides up to date information on Drought Monitoring and Reporting, Regional Crop Condition Reports, Surface Water Reports, and helpful links to the Agriculture Drought Risk Management Plan of Alberta (AARD 2013).

The federal and Alberta governments are sources of online information for municipalities and entities undertaking emergency management planning. A model plan for municipalities had been posted by AEMA (2015).

6.5.4 *Managerial Instruments*

Alberta has a lengthy history of employing participatory planning processes in respect of water, climate change, and drought in Q3 and Q4. Although Alberta is at the headwaters, it is obliged to deliver water to Saskatchewan and Manitoba pursuant to the Boundary Waters Treaty. Alberta's Water Act integrates **water** planning with a consultative practice employed in 2003 when Alberta's Water for Life Strategy was created (AG 2013; AE 2008). Province-wide consultations have occurred with approximately 300 people (A7) and the strategy has been revisited and reassessed periodically over the last decade (AG 2013). The Strategy focuses on issues of quantity, quality, and conservation of water, all related to preparing for and during **drought**. The strategy initiated three important activities to address unstructured Q3 problems: (1) planning for future management of water via the provincial **Climate Change Adaptation Strategy**; (2) development of land-use frameworks; and (3) watershed planning through local watershed groups. A Water Management Plan for the SSRB (SSRP) has been established for the study area (AE 2006). Watershed Planning Advisory Councils also have source water protections plans in river basins in the study area, all evidencing strong Q4 activity.

Municipalities have emergency response plans and are entering into cooperation agreements with neighbouring municipalities in the event of emergency. Discussion surrounding regional emergency plans is occurring (A7).

The Agriculture Drought Risk Management Plan (ADRMP) sets out government action plans for response to **drought** (AARD 2010). This plan establishes linkages to other relevant policies and plans of other departments (such as climate change), outlines activities for drought preparedness, monitoring, reporting and response. Alberta is just starting to develop local and regional drought plans (*ibid.*). Alberta also has insolvency instruments (see Sect. 5.5) to assist agricultural producers restructure after debilitating economic loss.

6.6 Adaptive Governance and Problem Structuring

Figure 6.2 summarizes the instruments relating to policy framing of climate change, d&f, and the relevant environmental governance approach. Although actors do not generally frame instruments in relation to climate change or unstructured Q3 problems, all quadrants are populated (other than Q2).

Unlike Saskatchewan, there is a policy instrument to deal with the unstructured Q3 problem of **climate change** mitigation and adaptation; however, the instruments (carbon offset market, etc.) reside in Q1 as they are managed in a technocratic manner with no recent input or participation of people. Planning for drought (the moderately structured problem) is performed by local watershed groups in Q3. Water management plans, water transfers, and management of irrigation districts are all

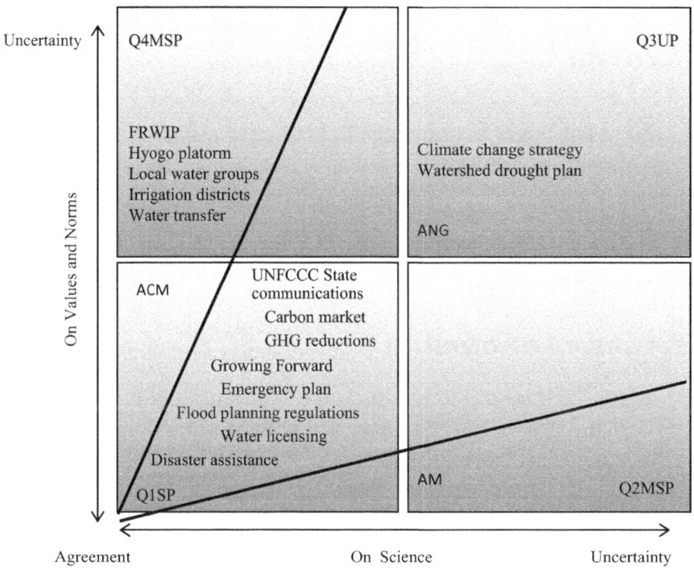

Fig. 6.2 Alberta's instruments and problem structuring
ACM Adaptive co-management, *Q4ACM* Quadrant 4 moderately structured problem, *ANG* Anticipatory governance, *Q3UP* Quadrant 3 unstructured problem, *Q2MSP* Quadrant 2 moderately structured problem, *AM* Adaptive management, *Q1SP* Quadrant 1 structured problem

met in a moderately structured problem frame of Q4 through stakeholder consultation. As in Saskatchewan, a suite of economic instruments assists producers in respect of the structured problem of responding to d&f (Q1) and the FSP assists in measures to build resilience to d&f (Q1).

Interviewees confirmed that instruments are not framed in Alberta as facilitating adaptation to climate change in Q3. In fact, one government interviewee discussing the Irrigation Efficiency programme "didn't know that the link to adaptation to climate change had ever been made" even though one programme improved water use efficiency and reduced energy consumption (A3). Instead, programmes are framed in Q1 and aim to reduce risk in low water years and achieve water security. The suite of instruments for flood is fragmented in Q1 because flood zone mapping, building codes, and disaster assistance payments are not coordinated. Further, these instruments only respond to flood and don't anticipate the driver of increasing flood due to climate change, a Q3 problem.

Although there are many examples of Q4 participatory measures (the Water Strategy, SSRP, ADRMP, integrated land use plans etc.), these plans are fragmented and not coordinated. Thus, the opportunity to deal with unstructured problems such as climate change and adaptation (Q3) and achieve social learning is lost. No incidences of adaptive management or Q2 moderately structured problems were discovered.

6.7 Impact of Instruments on Actors Measureby Mandate Effectiveness

Although Alberta has many instruments, the effectiveness of management and participation instruments is questionable because of fragmented implementation. This section analyses the impact of instruments, given drivers and grouped by function (see Sect. 3.3.5).

6.7.1 Regulatory Instruments

Interviewees did not regard Alberta's **climate change** regulations as effective; secondary sources confirm this. A reduction in GHG emissions has not occurred and is unlikely, given the increase in oil and gas pipelines (see CBC News 2014). Further: 1) the non-profit organization created to manage Alberta's carbon offset and energy efficiency rebate programmes was shut down in 2014 (Kleiss 2014); funds from carbon emissions are targeted at economic growth, not carbon reduction (see Hussain 2014: D2; ConBC 2015).

The regulation of water through license has responded well to **drought** in conjunction with the ability to transfer water interests (discussed below as an economic instrument). Civil servants enforce water licences through physical means (control mechanisms) in times of scarcity (A3). The regulatory licensing system and drinking water quality standards are both viewed by interviewees as effective.

The High River **flood** of 2013 illustrates the vulnerability created by the lack of requirement for regional emergency plans. 28 local declarations of emergency existed and many communities, such as High River were completely flooded and couldn't find, let alone implement, their disaster plan (A2). The provincial EMA became involved and there was mass confusion surrounding who was in charge and what was happening (A8). This situation was worsened because laws do not as yet prohibit building in flood zones; they are still awaiting passage in the provincial legislature (A2). The building requirement that infrastructure withstand a 1/100 event of **flood** was believed to be significant for enhancing resilience, but requiring review; one interviewee (A6) questioned its efficiency.

6.7.2 Economic Instruments

Economic instruments have been successful in changing water practices in times of **drought** and responding to **flood**. The economic instruments (Growing Forward 2) stabilize producer income and respond to d&f. The regulatory water licence in combination with the limited water market allowed water transfer and sharing amongst irrigators and licensees to maximize production during extreme drought. Some producers sold their water interests to others allowing them to produce a crop (which would have been unavailable without the transfer ability (A5; Hurlbert 2009a, 2009c)).[5] This system has also allowed municipalities with increasing populations to increase their water allotment through acquisition of other licences (A3). A positive, but not significantly large, step in planning for drought is the Irrigation Efficiency programme (A3) which encourages the use of new efficient technology (A3).

Government flood assistance is seen as efficient (A1, A8). The High River **Flood** of 2013 qualified for Disaster Financial Assistance as a 'large-scale natural disaster' (PSC 2014) and achieved its mandate of compensating homeowners for 90% of losses (ibid.).

6.7.3 Suasive Instruments

The suasive instruments in Alberta are similar to those in Saskatchewan and have similar effectiveness at achieving their mandate (see Sect. 5.7.3).

6.7.4 Managerial Instruments

Management instruments in Alberta include many bottom up, consultative instruments on water, climate change **adaptation and drought,** but they are not framed as adaptation instruments. These consultations and plans are not integrated or interconnected (Water for Life, Alberta Drought plan, climate change, SSBWMP, integrated land use planning, etc.), being mostly fragmented and incoherent, and the local environment continues to degrade (Stewart 2011).

Flood related plans are poorly coordinated. They have different definitions of flood plains in different jurisdictions complicating advance planning and the federal government has started flood plain mapping on several occasions in the past without finishing it (C9). During floods, local people and civil society members viewed this poor coordination as problematic (A8, A1), pointing to the complexity of the multiple jurisdictions involved in the SSRB (A1). Natural disasters occur across municipal, provincial, and even international borders with added layers of complexity not accounted for by instruments (C9). Between local municipalities, irrigation districts, emergency response service providers, provincial governments, the RCMP, and Canadian military, complexity and communication problems were cited as disadvantages (A8).

[5]Approximately 70 licensees as well as the community of Lethbridge didn't agree to the sharing arrangement and received orders to stop water withdrawals (Hurlbert 2009a, c; A5).

6.8 Assessment of Learning

Table 6.3 assesses how the instruments contribute to social learning. Dryland farmers have experienced the same level of single loop learning as Saskatchewan facilitated by the same instruments (see Sect. 5.8).

Triple loop learning has historically occurred in Alberta. One of the most significant triple loop learnings occurred after the Great Depression (1920s–1930s) with the creation of the Special Areas of Alberta. The Special Areas Board was created to manage repossessed abandoned land. The Board leases this land by way of Crown lease and community pastures to the few remaining viable farmers and ranchers. Private ownership was replaced by public ownership with managerial control in the Board (Marchildon 2007). Despite two major reviews by the provincial government (1953 and 1960) this dictarial governance arrangement has remained and a more democratic form rejected (*ibid.*). This change in land ownership reflects a fundamental change in underlying world-views, power dynamics and values influencing land management.

Two historic examples of double loop learning exist. The first is the use of oil revenues to create the irrigation districts in Alberta in the early twentieth century to ensure long-term agricultural producer viability (A5). The second example is the water transfer within irrigation districts in the 2001 drought (discussed in Sect. 6.7.2), which implied a movement from the regulated Q1 water to a co-managed Q4 water resource (Morito 2008).

Learning in respect of **flood** is problematic. After every flood, an inquiry occurs, a report on how to better plan for and respond to flood is prepared, recommendations are made, and none are implemented (A1). High River has been flooded many times: 1995, 2005, 2008, 2011 and 2013, and yet federal government disaster assistance allows homeowners to rebuild in the flood area (A6). There is no flood

Table 6.3 D&F: Assessment of learning in Alberta

Level/learning evidenced	**Change in practice – single loop learning** (one way information flow – Q1)	**Change in assumptions or models – double loop learning** (two way information flow – Q2 Q4)	**Change in values, norms, worldviews or power dynamics – triple loop learning** (iterative information flow – Q3)
Individual	Changes in agricultural practises, such as minimum tilling	Adoption of irrigated technology Water transfers in 2001	Embracement and continued support of Special Areas Board
Regional	FRWIP and FSP changing practises, e.g. reverting drylands to grasslands	Historic policies and instruments advancing irrigation development	Creation of Special Areas Board
National	–	–	–

management plan for Alberta and even after the 2013 High River flood, a full review of the flood policy has not occurred (A6). Climate change hasn't been considered, even though all of the previous flood reports had recommended it (A6).

The Oldman River dam completed in 1991 demonstrates zero loop learning. After decades of dialogue with local First Nations, the provincial government deliberately selected the site to avoid co-management with the First Nations (Daschuck and Marchildon 2006) and excluded local participation in the selection, building, and management of the dam (Glenn 1999). The Oldman River dam is an attempt to manage a problem in a Q1 structured manner, with poor success, civil unrest and resulting placation, manipulation, and mistrust (Glenn 1999).

Although there are many instruments in Q3, many flood instruments have questionable effectiveness and there is little triple loop learning (see Fig. 2.4). However, significant robust examples of iterative information flow, participation, and the existence of Q4 policies show promise, if the overarching value of economic growth can make room for the value of carbon reduction (which is not necessarily a dialectically opposing goal).

6.9 Instrument Impacts on Livelihood Capitals

Human health and social capital are negatively affected in times of d&f, while instruments help large producers gain economic capital and more vulnerable ones to ironically transition into livelihoods in the oil economy (see Table 6.4).

6.9.1 *Human Capital*

Both d&f have had negative impacts on health. Drought increases psychological distress (Fletcher and Knuttila Fletcher and Knutila 2016) with no instrument addressing it (Kubik and Moore 2005). The High River Flood of 2013 resulted in deaths due to heart attack and worsening health, especially in those already suffering poor health (A8, A1). Psychosocial counselling is required after an emergency (A1, S7, S4), but is prevented by the driver of government atrophy (A1).

6.9.2 *Social Capital*

Increasing urbanization and changing farm size has reduced the number of people participating in rural social networks (Fletcher and Knuttilla 2016) and added strain to infrastructure funding, health, education, and support services (A7).

Emergency planning, or the 'incident command' system, is reducing social capital. In incident command, one department takes control of the area and is in charge

Table 6.4 Assessment of main instruments

	Instrument	Effect	+ Impact on Capitals	(−) Impact on Capitals
Reg	Environ. and Drainage liability (civil claim)	–		– Economic
	Emergency measures planning requirements	+	+ Human/Social	
	GHG reductions	–		– Natural
But miss	Volumetric water charge and return stream flow calculations			
	Flood zone differentiation			
	Flood zone relocation			
Econ	FRWIP	+	+ Technology	
	Loan instruments	+	+Economic	
	Irrigation efficiency	+	+ Economic	
	Tradable (sharing) water rights	+	+ Economic	
	Federal and provincial disaster assistance	+	+ Economic	
	Growing Forward – Agri-programs	+	+ Economic	
	Carbon Offset Market	–		– Natural
But miss	Private flood insurance			
	Financial assistance for renters, small business, employees			
	Financial assistance for communities planning and training in emergency response			
	Financial assistance for adaptive measures (i.e. irrigation)			
	Irrigation Expansion			
Suas	FSP	+	+ Natural	=
	Municipal emergency plans	+	+ Human	
But miss	Long term mental health services for victims of flood			
	Resilience disaster planning			
	Participation in UNCCD			
Management	Local watershed governance	+	+ Social	
	Irrigation association constitutions	+	+ Social	
	Cooperation agreements for disaster recovery	+	+ Economic	– Economic/Natural
	Insolvency management	+	+ Economic	

(continued)

Table 6.4 (continued)

	Instrument	Effect	+ Impact on Capitals	(−) Impact on Capitals
But miss	Long term water management plans on integrated basis; integrated climate change and adaptation in plans			
	Flood plan. Regional emergency plans			
	Demand management water			
	Coordinated planning for both d&f (infrastructure)			
	Online coordinate support and service for local communities			
	Inclusive, participatory development instruments			

Inst Instrument, *Reg* Regulatory, *Miss* Missing, *Econ* Economic, *Suas* Suasive, *Mgt* Managerial
++ Effective; + moderately effective; – Ineffective

of commanding other departments, NGOs and CSOs when a disaster occurs. They exclude public participation (contrary to the international triple loop learning that people should participate throughout disaster risk reduction planning; see Sect. 4.6). Emergency planning institutions do not thus far promote resilience in disaster response and preparation (A1).

Several instruments have facilitated social capital:

(a) A historic policy encouraged irrigation development (A4) by funding the cost of irrigation infrastructure which led to the current irrigation associations, value added food processing industry, and supported NGOs, CSOs and private businesses;
(b) Management instruments facilitating source water protection planning have advanced integrative local planning (AE 2003, 2008). Local groups have strengthened friendships/familial ties and engaged people in discussions to initiate adaptive responses to drought by appealing to senses of honour and neighbourliness (Corkal et al. 2016);
(c) Regulatory instruments requiring local communities to respond to emergencies have encouraged informal networks of producers and community members to respond to fire, water shortages and blizzards (Hurlbert et al. 2015);
(d) The economic instrument of water sharing contributed to transferring of water interests in 2001 through informal sharing arrangements (Morito 2008) instead of pursuing formal legal water rights and interests. This instrument allowed people to ‘do the right thing’ solving the water shortage between themselves (A5).

6.9.3 Economic Capital

Many producers have diversified incomes beyond the farm with off-farm income. Ironically, the oil and gas industry offers high paid employment to supplement farm income (Wandel et al. 2010). This allows many to continue farming, but doesn’t negate the driver of the growing inequality between large and small agricultural producers (see Sect. 6.3).

As in Saskatchewan, (1) economic instruments assist agricultural producers to adapt by stabilizing income in times of d&f, (2) insolvency instruments help agricultural producers to transition to alternative livelihoods, and (3) many producers successfully utilize economic instruments including credit to expand operations and access technological capital. However, these instruments will be inadequate for the anticipated longer droughts lasting from 3 to 5 years (A5). Eventually producers may not be able to continue in business and a ripple effect might occur throughout the Southern Alberta economy with the closure of food processing plants and loss of employment (*ibid.*).

In addition to inequality in relation to drought adaptation, there is also inequality in relation to flood compensation. Disaster assistance applies inequitably; homeowners and municipalities receive compensation and can rebuild; but vulnerable people renting accommodations and business owners received nothing during the 2013 flood. They couldn't access disaster funds to rebuild their homes and businesses; employees lost their jobs (A8). Conversely, high-income people continue to reside in the flood plain, and in Calgary (the capital city) high-income people can afford to fortify their properties with new pump systems, more resilient building materials, and even metre-high concrete barriers around the yard (Markusoff 2014).

6.9.4 Technological Capital

Unequal economic capital affects access to technological capital. Although the irrigation efficiency instrument is available to all, large producers access economic instruments and technological capital more extensively than small producers (C5). Alberta has a historical, extensive irrigation capital since the early twentieth Century, but no new irrigation districts have been built in decades (A3).

Instruments fail to link water retention projects for drought, with those to prevent flooding and this negatively impacts technological capital. Institutions for flood management are separated from those on drought management and poorly integrated (C5, A3). Planning in 2015–2016 is just starting to link these issues together.

6.9.5 Natural Capital

Alberta has a bountiful supply of natural resources that fuel its economy and provide occupations and revenue sources for its residents. Climate change may positively impact on agriculture (Adams et al. 2001) by extending growing seasons and warmer temperatures (Pfeifer and Habeck 2002). However, oil sands, their related pipelines, and natural gas fracking challenge the local to global ecosystem (Birn and Khanna 2010).

Economic instruments (FRWIP – assisting with building water infrastructure) and suasive instruments (FSP – assisting with environmental farm practices) bolster the natural capital of producers and the Special Areas Board (see Sect. 6.5) promotes resilient agricultural practices. However, these instruments haven't yet fully addressed the driver of deteriorating ecological services (FPTGC 2010).

6.10 Re-designing Instruments with the ACW

This chapter explored how Alberta's institutional policy framework on climate change, d&f impacts capitals of agricultural producers given local and national drivers. I now assess the adaptive capacity of Alberta (see Fig. 6.3) and make recommendations for policy redesign to improve the adaptive capacity of the governance institutions.

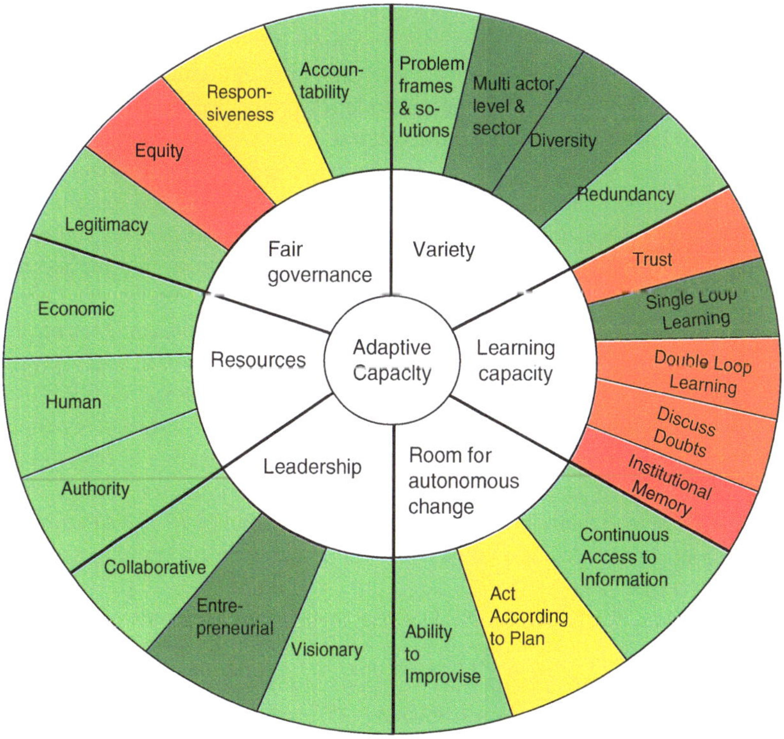

Green	Light Green	Yellow	Light Orange	Red
Very High	High	Medium	Low	Very Low

Fig. 6.3 Alberta's ACW

Starting with '**variety**,' and working clockwise direction on the ACW, Alberta appears to have many instruments and policies responding to climate change, d&f. Overall, Alberta embraces the risk of the unstructured policy problems of climate change (Q3), and the sub policy problems of d&f.

The development of irrigated agriculture demonstrates historic double loop **learning** and the creation of the Special Areas, triple loop learning. Both examples facilitated producer adaptation to climate variability and change and required leadership and financial contribution from the government. It is not clear that this leadership still exists; the financial instruments do not.

Dry land producers and ranchers have been as innovative and adaptive as their Saskatchewan counterparts improving environmental practices and the profitability of their operations through innovative technological improvements and single loop learning with **room for autonomous change**. The recent High River flood accentuated deficits in learning and despite repeated floods and government reviews, recommendations do not get implemented; and the building codes, insurance, flood plain building restrictions, and defects in linking infrastructure to problems of both d&f, and regional emergency planning are poorly integrated and incoherent. There is thus only partial double loop learning and inadequate room to discuss doubts, reflected in the light orange colour in Fig. 6.3.

The water governance system illustrates **autonomous change** (with the ability to transfer) **and leadership** (civil society water planning, entrepreneurial producers, collaborative irrigators). However, the multiple, disconnected, singular, participation instruments have resulted in dwindling **trust**.

Resources appear sufficient, except in relation to small producers, renters and small businesses. The economic and human resources of the actors are substantial and government exercises legitimate authority. However, small farmers and renters tend to be ignored in inequitable instruments (as has been illustrated in the recent High River Flood).

Federal personnel interviewed were concerned about the instruments on d&f. First, a heightened 'security' focus on terrorism rather than natural to anthropogenic causes has led to top down, fragmented and disconnected approaches minimizing participation, resilience and inclusive development (C7). Privacy, openness, accountability, and **fair governance** suffered negatively impacting access to information and autonomous change. This was worsened by too much focus on critical top-down incident management and not community resilience (C2) that empowers decision making at lower levels using a cluster approach to bring people together for bottom up planning (C2).

Redesign options for Alberta include strengthening the ineffective instruments, addressing all drivers, and focusing on constrained livelihood capitals identified in Table 6.5. Issues of effectiveness arise in relation to the tools addressing the unstructured problem of **climate change mitigation** (Q3), perhaps due to sporadic participation instruments without climate adaptation projects or plans and inadequate tackling of the driver of climate denial. Climate denial is also illustrated by people migrating to urban municipalities and jobs within the natural resource sector of oil, gas and oil sands, a development not conducive to addressing climate change. It is

Table 6.5 Redesign options for Alberta

Analysis unit	Description
Ineffective instruments	Climate mitigation, carbon offset market, climate change, water plans
	drainage permits, liability, fines, flood zone mapping, building standards, environmental liability, municipal zoning bylaws
	National Disaster Mitigation Strategy, National Hyogo Platform for DRR
Missing instruments	Volumetric water charge, return stream flow
	Private flood insurance, missing financial disaster assistance for renters, etc., long term mental health care, resilience disaster planning
	Participation in UNCCD
	Irrigation expansion incentive
	Inclusive, participatory development instruments, local water planning
Unaddressed drivers	Government austerity
	Climate change, climate denial
	Growing inequality
	Deteriorating ecosystems
Weak adaptive capacity dimension	Equity
	Variety (problem frames and solution, redundancy)
	Learning capacity (discuss doubts)
Constrained livelihood capital	Health (mental health after disaster)
	Natural

clear that not utilizing adaptive management and anticipatory governance inhibits adaptive governance in Alberta. The institutional governance framework surrounding d&f in Alberta suffers from a lack of interconnection, interconnection of instruments with people, nature (or natural capital), and climate change.

To improve adaptive governance, the most important driver that needs to be addressed is climate change and the deterioration of ecosystem services reducing producers' natural capital. Current instruments are inadequate. Focus on learning through measuring GHG reductions and implementing adaptive management techniques is required. Missing instruments should be implemented including the UNCCD, international best practice instruments of adaptation and disaster resilience, and irrigation expansion. Special attention to growing inequality, the institutional dimension of inequity, and the lack of triple loop learning should be paid. As in the case of Saskatchewan, building an adaptive governance institutional system will require more focused attention on long term and transitionary time frames and the interconnections of people, capitals, and instruments to take into account the drivers or causes of problems – most importantly that of climate change.

References

AARD (2010) Agriculture drought risk management plan for Alberta – 2010. Last reviewed and revised July 12, 2010. Available at: www1.agric.gov.ab.ca./$department/deptdocs/nsf/all/ppe3883#actions. Accessed 13 Sept 2014

AARD (2013) Drought monitoring and reporting. Last reviewed and revised May 14, 2013. Available at: www1.agric.gov.ab.ca./$department/deptdocs/nsf/all/ppe9019. Accessed 13 Sept 2014

AARD (Alberta Agriculture and Rural Development) (2015a) Growing forward 2 Pro-grams Irrigation Efficiency. Available at: www.growingforward.alberta.ca/Programs?index.htm?contentId=IRRIG_EFF_PRG&useSecondary+true&active+yes. Accessed 31 Mar 2015

AARD (Alberta Agriculture and Rural Development) (2015b) Agroclimatic information service (ACIS). Available at: www.agriculture.alberta.ca./acis/AARD. Accessed 31 Mar 2015

Adams RM, Hurd BH, Reilly J (2001) Impacts on the US agricultural sector. PEW report. Climate change: science, strategies and solutions, pp 47–64

AE (2006) Approved water management plan for the South Saskatchewan River Basin (Alberta). Available at: http://environment.alberta.ca/documents/SSrb_Plan_Phase2.pdf. Accessed 29 Sept 2014

AE (2008) Water for life: a renewal. ISBN 978–0–7785-7670-9 November, 2008. Available at: http://environment.gov.ab.ca/info/library/8035.pdf. Accessed 30 Jan 2015

AE (Alberta Environment) (2003) Water for life. Alberta's strategy for sustainability. November, 2003. ISBN No. 0–7785–3058-2 Pub. No. 1/955. Available at: http://environment.gov.ab.ca/info/library/6190.pdf. Accessed 30 Jan 2015

AEMA (Alberta Emergency Management Agency) (2015) Alberta municipal affairs, milenet. Available at: www.milenet.ca/milenet/system/dsp_login.cfm

AESRD (2015) Alberta environment and sustainable resource development. Available at: http://esrd.alberta.ca/. Accessed 30 Apr 2015

AG (Alberta Government) (2010) AFSC, insurance, lending, income stabilization. 2009–2010 Annual Report. Available at: https://www.afsc.ca/doc.aspx?id=3857. Accessed 29 Sept 2014

AG (2013) Water for life. Available at: http://www.waterforlife.alberta.ca/. Accessed 13 Sept 2014

Barrow E (2010) Climate change scenarios for the Prairie Provinces. In: Sauchyn D, Diaz H, Kulshreshtha S (eds) The new normal. Regina, Canadian Plains Research Center

Birn K, Khanna P (2010) A discussion paper on the oil sands: challenges and opportunities. World Energy Congress. http://89.206.150.89/documents/congresspapers/403.pdf (01/04/2014). Accessed 19 Sept 2014

Bruneau J, Corkal DR, Pietroniro E, Toth B, Van der Kamp G (2009) Human activities and water use in the South Saskatchewan River Basin. Prairie Forum 34(1):129–152

Byrne J, Kienzle S, Sauchyn D (2010) Prairies water and climate change. In: Sauchyn D, Diaz H, Kulshreshtha S (eds) The new normal. Regina, Canadian Plains Research Center

CAUT (2013) A spotlight on the many ills that plague scientific pursuit in Canada CAUT Bulletin 60(10) December 2013.p 1. Ottawa

ConBC (Conference Board of Canada) (2015) Investing in GHG emissions-reduction technolog. Assessing the Economic Impact. Briefing February 2015. Ottawa

CBC News (2014) Pipeline from oilsands to Artic is feasible: Alberta study. CBCnews. Available at: http://www.cbc.ca/news/canada/north/pipeline-from-oilsands-to-arctic-is-feasible-alberta-study-1.2757637. Accessed 19 Sept 2014

Corkal D, Morito B, Rojas A (2016) Chapter 11: Values analysis as a decision-support tool to manage vulnerability and adaptation to drought. In: Diaz H, Warren J, Hurlbert M (eds) Vulnerability and adaptation to drought: the Canadian Prairies and South America, ISSN 1919–7144. University of Calgary Press, Calgary, pp 251–278

Daschuck J, Marchildon G (2006) Historical chronology of the Oldman River dam conflict. Research Report. IACC Project. Available at: http://www.parc.ca/mcri/pdfs/papers/iacc036.pdf. Accessed 13 Sept 2014

EC (Environment Canada) (2012). Backgrounder. reduction of carbon dioxide emissions from coal-fired generation of electricity regulation. http://www.ec.gc.ca/default.asp?lang=En&n=5C4438BC-1&news=D375183E-0016-4145-A20B-272BDB94580A. Accessed 2 Jan 2013

FPTGC (Federal, Provincial and Territorial Governments of Canada) (2010) Canadian biodiversity: ecosystem status and trends 2010. Canadian Councils of Resource Ministers. Ottawa, ON. vi + p 142

Fletcher AJ, Knutila E (2016) Chapter 7: Gendering change: Canadian farm women respond to drought. In: Diaz H, Warren J, Hurlbert M (eds) Vulnerability and adaptation to drought: the Canadian Prairies and South America, ISSN 1919–7144. University of Calgary Press, Calgary, pp 159–178

Galloway G (2014) Purge of Canada's fisheries libraries a historic loss, scientists say The Globe and Mail, Wednesday, January 8, 2014. p A8. Available at: http://www.theglobeandmail.com/news/politics/purge-of-canadas-fisheries-libraries-a-historic-loss-scientists-say/article16237051/. Accessed 13 Sept 2014

Glenn J (1999) Once upon an oldman, special interest politics and the oldman Riveer am. UBC Press, Vancouver

Graveland B (2013) Brazilian company buys XL foods plant involved in E. Coli meat recall. Macleans, January 9, 2013

Harvey D (2005) A brief history of Neoliberalism. Oxford University Press, Oxford/New York

Hurlbert M (2009a) Integrating climate change adaptation into the law. Retfaerd Argang 32(3):23–39

Hurlbert M (2009b) Comparative water governance in the four western provinces. Prairie Forum (a special edition on climate change) 34(1):181–207

Hurlbert M (2009c) The adaptation of water law to climate change. Int J Clim Chang Strat Manag 1(3):230–240

Hurlbert M, Andrews E, Tesfamariam Y, Warren J (2015) Governing water: deliberative institutions and adaptation. Research Report (157 pp). Available at: www.parc.ca. Accessed 13 Sept 2014

Hussain Y (2014) Oil sands titans see promise in rising carbon emissions. The Leader-Post, The Financial Post, Friday May 2, 2014, p D2. Available at: http://www.leaderpost.com/business/sands+titans+promise+rising+carbon+emissions/9799174/story.html. Accessed 15 Sept 2014

IPCC (2014) Summary for policymakers. In: Edenhofer O, Pichs-Madruga R, Sokona Y, Farahani E, Kadner S, Seyboth K, Adler A, Baum I, Brunner S, Eickemeier P, Kriemann B, Savolainen J, Schlömer S, von Stechow C, Zwickel T, Minx JC (eds) Climate change 2014, mitigation of climate change. contribution of working group iii to the fifth assessment report of the intergovernmental panel on climate change. Cambridge University Press, Cambridge/New York

Johnstone B (2012) Farms getting bigger, but farmers fewer and older, the Leader-Post, May 11, 2012, p C1

Kleiss K (2014) Alberta climate change non-profit folds two years after government pulled funding. Edmonton Journal. June 3, 2014. Available at: http://www.edmontonjournal.com/Alberta+climate+change+profit+folds+years+after+government+pulled+funding/9899934/story.html. Accessed 13 Sept 2014

Kubik W, Moore RJ (2005) Health and well-being of farm women: contradictory roles in the contemporary economy. J Agric Saf Health 11:249–256

Markusoff J (2014) Metal flood wall protects Calgary home, The Sunday Post, the leaderpost.com, Sunday, June 1, 2014, p A3. Available at: http://www2.canada.com/calgaryherald/news/city/story.html?id=9ae7a48c-071d-4b7a-b88d-2580b4ea3b33. Accessed 13 Apr 2015

Marchildon GP (2007) Institutional adaptation to drought and the special areas of Alberta. Prairie Forum. 32(2)

Morito B (2008) Institutional values and adaptive capacity to climate change, 2008. Institutional Adaptation to Climate Change Research Project, University of Regina. http://www.parc.ca/mcri/pdfs/papers/iacc085.pdf. Accessed 13 Sept 2014

Percy D (2004) The limits of Western Canadian water allocation law. J Environ Law Pract 14(2004):315

Pfeifer RA, Habeck M (2002) Farm level economic impacts of climate change, 159–78. In: Doering O, Randolph J, Southworth J, Pfiefer R (eds) Effects of climate change and variability on agricultural production systems. Kluwer Academic Publishers, Boston

PPWB (Prairie Provinces Water Board) (2003) The 1969 master agreement on apportionment and by-laws, rules and Procedures. Prairie Provinces Water Board, Regina

PSC (Public Safety Canada) (2014) Disaster financial assistance arrangements (DFAA). Available at: www.publicsafety.gc.ca/cnt/mrgnc-mngmnt/rcvr-dsstrs/dsstr-fnnd-ssstnc-rrngmnts/index-eng.aspx. Accessed 13 Sept 2014

Remax (2014) Farm report 2014. Available at: http://download.remax.ca/PR/FarmReport/FinalReport.pdf. Accessed 16 Apr 2015

Rocher E (2013) Climate change played a role in high river flood. Western Wheel. August 21, 2013. Available at: http://www.westernwheel.com/Climate-change-played-a-role-in-HIgh-River-flood-20130821. Accessed 30 Mar 2016

Sauchyn D, Kulshreshtha S (2010) Chapter 17: agricultural adaptation through irrigation. In: Diaz H, Warren J (eds) Vulnerability and adaptation to drought in the Canadian Prairies. University of Calgary Press, Calgary

Statistics Canada (2011) Farm debt outstanding: agriculture economic statistics. Minister of Industry, Ottawa, 21–014-X, vol 10, no.2. Available: http://www.statcan/gc/ca/pub/21-014-x/21-014-x2011002-eng.pdf. Accessed 13 Sept 2014

Statistics Canada (2012) Snapshot of Canadian agriculture: farm and farm operator data. Statistics Canada, Ottawa, 95–640-XWE. Available at: http://www.statcan.gc.ca/pub/95-640-x/2012002/01. Accessed 13 Sept 2014

Stewart J (2011) Watershed management & cumulative effects: what private landowners need to know. Presentation to the 2011 national environment, energy and resource law summit, water and the law: allocation, trade, use and protection. Fairmont Banff Springs April 7–9, 2011

Wandel J, Pittman J, Prado S (2010) Rural vulnerability to climate change in the South Saskatchewan River Basin. In: Sauchyn D, Diaz H, Kulshreshtha S (eds) The new normal. The Canadian Prairies in a changing climate. CPRC Press, Regina, pp 245–258

Yalnizyan A (2013) Study of income inequality in canada – what can be done, Presentation to the House of Commons Standing committee on Finance. Available at: https://www.policyalternatives.ca/publications/reports/study-income-inequality-canada%E2%80%94what-can-be-done#sthash.Q1YYvuNR.dpuf. Accessed 14 Apr 2015

Laws

Passenger Automobiles and Light Trucks, Heavy-Duty Vehicles, Renewable Fuels and Reduction of CO2 Regulations pursuant to the Canadian Environmental Protection Act, 1999. SC 2004. c. 15, s. 31

Reduction of Carbon Dioxide Emissions from Coal-fired Generation of Electricity Regulations, P.C. 2012–1060 August 30, 2012 pursuant to the Canadian Environmental Protection Act, 1999. SC 2004, c. 15, s. 31

The Alberta Land Stewardship Act. SA 2009. C. A-26.8

The Bankruptcy Act. RSC 1985. C. B-3

The Boundary Water Treaty of 1909

The Canadian Environmental Assessment Act. SC 1992, C. C-11

The Canadian Environmental Protection Act, RSC 1985, C. 16 (4th Supp)

The Climate Change Emissions and Management Act. SA 2003, C.C-16.7
The Constitution Act. 1982, being Schedule B to the Canada Act, 1982, 1982, C 11
The Emergency Management Act. RSA. c. E-6.8
The Environmental Protection and Enhancement Act. RSA 2000, C. E-12
The International Boundary Waters Treaty Act. RSC. 1985, C. I-17
The Water Act. RSA 2000. c. W-15

Chapter 7
Case Study Coquimbo, Chile

7.1 Introduction

This chapter explores the institutional policy framework of Coquimbo, Chile of climate change, d&f and its impacts on agricultural producer capitals, given local, international, and national drivers. The chapter is structured like all case study chapters.

This chapter shows that water and irrigation management instruments have been highly successful in Chile when combined with the private water market and the driver of increasing trade liberalization. Double loop learning has resulted with the emergence of a large agricultural export economy. A strong central Chilean government has capitalized on the resilience and inventiveness of very large agricultural producers connected multi nationally supporting this agriculture with access to large infrastructure (such as dams), issuance of water licenses, and agricultural technology, including irrigation. The growth of the export economy has benefited large producers due to the combination of trade liberalization and neo-colonialism. However, strength overdone may also be weakness.

Inequity within the governance system is a concern. Ineffective and sparse regulatory instruments allow powerful water rights holders like those in mining, energy, and export agriculture to usurp community drinking water and traditional local water practices predating the 1981 Chilean constitutional privatization of water. Despite economic instruments targeting small agricultural producers, there is growing inequality; Chile's participatory instruments are not working. Effective climate change instruments are missing. The past 7 years of emergency drought declarations in the study region (Co7) and resulting reductions in water allocations for agricultural producers have created a 'new normal' that constrains many livelihoods.

M.A. Hurlbert, *Adaptive Governance of Disaster*, Water Governance - Concepts, Methods, and Practice, DOI 10.1007/978-3-319-57801-9_7

7.2 The Case Study Area

The study region in Chile is within Region IV, the Elqui River Basin. Located within the Coquimbo Region in the near north region of Chile, it is a semi-arid, dry land containing both desert (the Atacama) and fertile valley (which characterizes central Chile) (Pizarro 2009). The urban district of La Serena and rural communities of Vicuna and Pahuano are included (Salas et al. 2012). The dry season can be from 10 to 12 months and can have a water deficit of approximately 700 mm at the coast and 1500 mm between the coast and the Andes Mountains (which separate Chile from Argentina). High snow accumulation in the Andes (including ten glaciers; Valdez-Pineda et al. 2014) melts and then feeds the rivers that proceed through three transverse valleys emptying into the Pacific.

The entire region has 687,806 people with almost half residing in the city of La Serena on the coast (GC 2010). The Andes range runs close to the sea making this region very mountainous. Mining and agricultural activities predominate. Within the valleys many workers are required for producing fruit, vegetables, grapes for wine and '*pisco*' (local brandy) and flowers on irrigated land. Dryland between the valleys is significant and most of the regional rural poor population inhabits this area (Kubik et al. 2010). The main production in the dry land areas is cattle and goats, small-scale agricultural production, traditional fishing and mining (ibid.). The irrigated land area represents only 1.8% of the watershed while the dryland area is 72.3% (Salas et al. 2012) (Fig. 7.1).

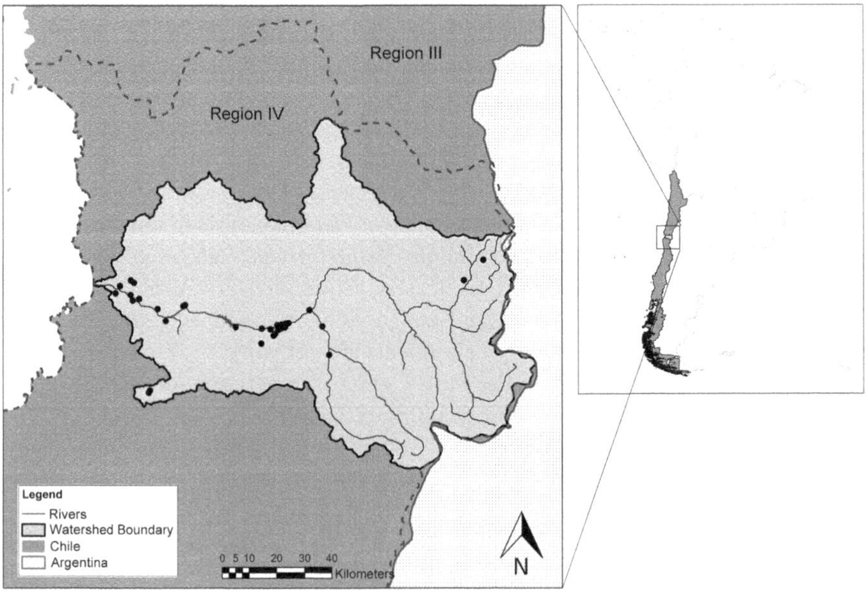

Fig. 7.1 Study region in Coquimbo, Chile

7.3 Main Drivers Impacting Rural Agricultural Producers

In addition to climate change, increasing trade liberalization combines with colonial elitism to augment the growing inequality and bifurcates the legal system privileging formal law over water custom. These drivers are summarized in Table 7.1.

Coquimbo has already experienced warmer temperatures, less rainfall, and retreating glaciers (Valdez-Pineda et al. 2014). Increasing drought, reduced streamflow, and increasing frequency of extreme events are anticipated in the future (OECD 2013; Hannah et al. 2013). Available water discharge may decrease by 20–35% by 2050 and the mean climatic suitability for viticulture in Chile may decrease by up to 25% (Hannah et al. 2013). Water deficits (situations where water demands exceed supply) are expected to increase dramatically in the future and most significantly in the study region (Valdez-Pineda et al. 2014). There will be an increase in drought and river overflows due to El Nino southern oscillation events (OECD 2013). This means decreases in ecosystem services of surface and ground water with significant impact on the quality of water resources and a reduced ability to dilute and regulate pollutants and liquid waste.

Chile has prioritized the creation of an export economy of food production through the expansion of irrigation capacity (Reyes et al. 2009), leveraging the driver of trade liberalization. In the last three decades Chile's real GDP growth was 6.2% per annum, most of which is based on activities requiring increasing amounts of water (Valdez-Pineda et al. 2014; Brown and Pena 2003). Agriculture has boomed with significant investment by agri-business in large-scale high-value irrigated crops like fruits, vegetables, grapes, and grapes for wine and *Pisco*, and value-added processing (FAO 2003a). Wine exports contributed over USD 1.4 billion to the economy in 2012 (ODEPA 2013). Business is heavily involved in water development. A user-pay, full water cost mentality has developed and water is not regarded as a human right (Larrain 2014). As production units get larger, people move to cities or seek mining jobs; labour is scarce (Hadarits et al. 2016; Hill 2013). Depending on wine exports, producers are affected by the value of the USD (Hadarits et al. 2016). Large farmers are impacted by inflation, currency instability and access to credit for investment in new technology (Salas et al. 2012).

Trade liberalization impacts Chile in the context of colonial elitism, low state involvement (Budds 2004) and historical dictatorship (Carruthers 2001) exacerbating inequality as barely 20% of Chileans have incomes on par with those of a developed country; most earn significantly less (see World Bank 2009). The political elite benefit while small agricultural producers and socio-ecological resilience are negatively impacted (Clarvis and Allan 2013).

Formal law is disconnected from customary practice (Dourojeanni 2014). In spite of the formal water market, the *de facto* water allocation is often based both on historical hydrological conditions as well as agreements (Vicuna and Meza 2012). The Juntas deliver the water and the Irrigation Associations distribute it often based on rules and practices dating back to 1908. Internal conflicts are handled within the system, partly because these conflicts need to be solved immediately (Peralta 2014).

Table 7.1 Drivers impacting agricultural producers' capitals

	Driver – direct		Driver – indirect	
Category	Local	National	Local	National
Demographic	Shortage of farm labour (Salas et al. 2012)		Urbanization/population growth	Culture of neoliberal market with colonial elitism (Budds 2004; Carruthers 2001)
	Changing size of farms		Aging producers (Salas et al. 2012)	Lack of transparency and enforcement of water law
Economic	Active mining industry competes for labour (Salas et al. 2012)	User pay pricing of water	Aging infrastructure (roads, dams)	USD value impacts the export economy (Salas et al. 2012)
				World recession 2008
				Reduced government size and budget
Political Economy	Increasing cost of inputs (Hadarits et al. 2016)	Export food economy created through irrigation (Reyes et al. 2009)	–	Increasing trade liberalization (Reyes et al. 2009)
				Growing inequality
				Social support left to private sector (ibid.)
				Increasing demand for energy
Social	Informal water practices trump legal rules (Gobernanza del agua 2014)	–	–	Priority of economy over environment
Natural	Climate change Loss of ecosystem services	–	–	–

The system is the 'law of the jungle' where formal water law is avoided (Gobernanza del agua 2014). Environmental laws aren't enforced, environmental impact assessments aren't conducted, and illegal extractions of groundwater aren't policed (Hill 2013). Often legal rules are subservient to an individual or company's personal bargaining power (Bauer 1997; Carruthers 2001).

7.4 Institutions (Organizations) That Build Capacity for Climate Change, d&f

Chile is a democratic republic with a strong central government in Santiago. The President appoints governors to administer the fifteen regions, the case study area being one. This section sums up the main formal institutions.

The major institutions include the National Environmental Commission (NEC), the Ministry of Environment (MOE) (both previously CONAMA) and its **climate change** office and the National Institute for Agriculture Development (INDAP). The federal Chilean General Directorate of Water (DGA) governs **Water** providing hydrological expertise on river flood and flow, and aquifers (DGA 2007). Other entities develop resources and provide social support, but there is a strong influence of market-led policies (Reyes et al. 2009: 6). The Hydraulic Works Authority (DOH) does research and plans water infrastructure with DGA's input. The Inter Ministerial 'National Irrigation Commission' (CNR) involving DOH, DGA and five other Ministries improve or develop new irrigation programmes. The Superintendent of Sanitary Services (SISS) (Reyes et al. 2009) deals with drainage or contamination events affecting surface or ground water. The DGA and *Juntas de Vigilancia* manage water quantity in a fragmented manner (Water Code, Art 122 and 146) (Hill 2013).

Water is the responsibility of the central state government, not the provinces or regions (as in the other case studies) thereby reflecting a grouping of disassociated institutions in the study area. This is because the DGA lacks supremacy over other institutions (even mining and agricultural ministries; Romero et al. 2012), is short on budget and staff (Valdez-Pineda et al. 2014), and 'lacks power' (Hill 2013: 202). Power is consolidated in the central government that controls public funds and decision-making (Barton 2013); regional representatives appointed by the President are young and have no background in water. Citizen bodies and international NGOs attempt to protect the environment and social health of the country (Hill 2013) but are weak (Hill 2013: 250; Reyes 2014).[1]

In times of emergency **drought**, the DGA makes a declaration and the Juntas who oversee basin level water distribution implement it. **Floods** impact the study area through mudslides, worsened by construction of infrastructure such as roads

[1] Some contradictory opinions surround the relative strength and presence of NGOs and civil society organizations. Hill stated there is a "growing number and strength of NGO initiatives which serve to counter balance the pure economic interests" (Hill 2013: 205) and promote resilience.

and irrigation channels, deforestation, overgrazing, modified landscapes by large agricultural companies (Young et al. 2010), and unstable mine tailing sites (Cepeda 2008). While the National Emergency Office (ONEMI) organizes and coordinates technical plans for emergencies like earthquakes (which takes many resources (Reyes et al. 2009)), it does not deal with d&f. In 2008 a National Early Warning Center was created, but has not been used for flood warnings. ONEMI infrequently becomes involved in providing water (Co4). Local organizations respond to d&f by fulfilling their functions as best they can (ibid.): after a flood, rural potable water committees restore their system relying on interim municipal aid (*ibid.*); the Juntas and Irrigation Committees respond to drought by managing less water and private companies often provide public water and sanitation services.

Chilean water law requires water rights holders in each basin to form a water users association. Participation of users in election of directors and voting is directly proportional to the amount of rights held; thus larger users dominate the administration and decisions (Brown and Pena 2003). Other smaller groups include rural potable water committees (APR), agricultural communities, drainage communities, and groundwater communities. NGOs such as Chile Sustainable and Adapt Chile promote adaptation.

Private actors like large agri-business, mining companies and Endesa hold considerable power by owning water rights. This power is demonstrated in the hydroelectricity generation sector (Endesa included) reacting rapidly in the 1980s to purchase water rights, sometimes compromising whole river basins (Carruthers 2001).

7.5 Instruments Responding to Climate Change, d&f

The dominant instrument is the private water right, an economic instrument. In times of shortage an emergency declaration is made and a regulatory process of reductions ('*turno*') is employed proportionally reducing water allocations. Few instruments exist in relation to flood, or mudslides, although these events are common in the study area. The main instruments are summarized in Table 7.4.

7.5.1 Regulatory Instruments

Chile has few regulatory measures; most policy problems are left to be solved in the marketplace of the private sector. Recently, new measures to address water market shortcomings have been implemented. In 2005, regulatory water measures (a Q1 framing of problems) concerning ecological flows (in relation to new water rights), enforcement of allocations, and transfer provisions were introduced. To combat water hoarding, a levy for unused water rights was introduced (Law No. 20.099 of 2005). Fines and imprisonment were introduced for illegal water extraction and not

Table 7.2 Institutional water instruments

Principle	Description
Principle under which water is managed	Water is public good, however, the right of access is private property or marketable commodity (Article 19.24 1980 Chilean Constitution).
Allocation of water rights	None; Preferences are based on market rules after government initially allocates
Priorities	No priorities for different uses
Water Market	All water is governed by the market
Water pricing	One off fee for initial purchase of water rights then costs of distribution, operation and maintenance of the infrastructure according to the amount of rights (Hill 2013); water is priced in the market (Brown and Pena 2003).
Water allocation dispute resolution	The local *Juntas de Vigilancia* and the courts
Potable water accountability	The local level service providers and municipalities overseen by SISS
Governance accountability	The DGA is a government entity, *Juntas de Vigilancia* and Irrigation Associations are governed by members (water rights holders) in proportion to their rights

transferring water rights properly (OECD 2013). Drinking water monitoring and sanitation is regulated in a Q1 technocratic manner (SISS 2015).

Table 7.2 presents the 'principles' of water governance. Although CSOs promote the human right to water and local water committees assert presence in relation to the wicked problem of adaptation (Q3), no regulatory measures embrace these locally. These initiatives do not consider the unstructured problem of climate change (Q3).

On climate mitigation, Chile seeks to reduce its GHG emissions by 20% in relation to business-as-usual emissions in 2020 based on 2007 data (Nachmany et al., Nachmany et al. 2014), but not through regulatory instruments,[2] although one energy efficiency measure exists (but within a structured Q1 framing; Modification of the Electricity Law and Law 20.257 for Non-Conventional Renewable Energies 2008).

In times of **drought**, Article 314 of the Water Code provides for a Presidential Declaration of Drought Zones at the request of irrigators (Hill 2013) for a maximum of 6 months, if the river flow is less than 70% of the average. During this period, a traditional Spanish system of proportionally reducing the water of each rights holder through shifts managed by the Juntas –'*turno*' – is employed. Chile signed the UNCCD in 1996 endorsing a Zero Net Land Degradation target (UNCCD 2013); however, a national report hasn't been prepared yet.

Several regulatory measures encourage public participation including Environmental Impact Assessments and Public Participation law (Law N. 20,500).

[2] As a non-Annex I party (classified as mostly developing) of the UNFCCC, investment, insurance, and technology transfer are the desired options to meet Chile's special needs (UNFCCC 2015).

7.5.2 Economic Instruments

Economic instruments predominate in Chile especially in relation to water. The national Chilean Water Code (1981) ties water rights not to the land, but as marketable private property. The water market was created to expand irrigation and agricultural development. The Government initially sold water in a Q1 technocratic supply manner (followed by payments by rights holders for costs of distribution, operation, and maintenance of infrastructure based on the amount of rights held). After the initial government sale, the private water market reallocates water rights (Hill 2013) through sale, transfer, or inheritance as with any tradable commodity (Bauer 2004). Water rights holders don't have to state how they will use their rights, don't lose rights from non-use, and don't pay further taxes or fees to the government. In the study region all water rights have been allocated (Reyes et al. 2009). The state can only appropriate water rights through legislation and after providing compensation.

State funds stimulate large scale irrigation infrastructure and pay up to 75% of irrigation investment for small agricultural producers. CNR supervises irrigation investment in Chile through law 18,450 of 1985 (the Promotion of Private Investment in Irrigation and Drainage Works; Reyes et al. 2009; Brown and Pena 2003: 25). Irrigation is framed to enhance economic prosperity (Q1) and not to adapt to the unstructured problem of climate adaptation in Q3.

Other instruments include private insurance and government instruments aimed at small agricultural producers. Insurance includes private crop insurance (Reyes et al. 2009); micro-insurance (although it is not accessible by agricultural producers (McCord et al. 2013)); flood insurance for business loss, but not for homeowners); and catastrophic risk (generally underinsured in Latin America; e.g. the 2010 earthquake in Chile caused USD 30 billion in losses but only USD 8.2 billion of these losses were insured (Ernst and Young 2013)).

The government (INDAP) designs specialized programmes to build capacities of small farmers and peasant family agriculture that are framed in relation to the unstructured problem of climate change and **adaptation (Q4)**. A programme for local development-PRODESAL (INDAP 2015) provides services to about 50,000 families per year to respond to d&f and other climatic stressors including special crop irrigation incentives, soil recovery, and technology transfer programmes on farm management. INDAP has small emergency funds for supplies in times of drought or frost (USD 300 per family), a small irrigators initiative fund, a fund for goat herders, a fund for minor infrastructure repair or soil protection against desertification and erosion, and funds to enhance food processing and marketing (Reyes et al. 2009). Other instruments aimed at small agricultural producers address the Q3 problem of **climate change** (but none exist in the study area).[3]

[3] These include: a fund created in 2007 for local development projects that address climate change (Q3) (Fondo de Proteccion Ambiental (FPA)); funding and technical support provided by the GEF to develop and strengthen activities related to climate change (OECD 2013); and CDM initiatives (Barton 2013). Many capacity building initiatives also exist elsewhere in Chile (UNFCCC 2015; MAPS 2015).

7.5.3 *Suasive Instruments*

Many suasive instruments exist in relation to **climate change adaptation and water**. These instruments target wicked Q3 issues, but with little public participation and none within the study region. The Climate Change Office of MOE, has developed Eight National Adaptation Plans for key sectors including water resources, a National Plan on Climate Change Adaptation, and a Water Resources Adaptation Strategy. A National Climate Change Strategy was developed in 2006 (CONAMA 2006) and a National Climate Change Action Plan in 2008 (−2012) (OECD 2013; CONAMA 2008).

The DGA is responsible for a glacier protection policy that was developed in 2009 (OECD 2013) following the failure to develop a national glacier protection law because of a mining sector lobby (Reyes et al. 2009). The DOH has built a map of high **flood** risk zones and developed a plan for containment infrastructure to protect communities (ibid.) but has no implementation instruments.

The Water Resources National Strategy aims to create new **water** sources including reservoirs over the next decade and groundwater recharge projects (OECD 2011, 2013). The National Water Policy Report produced (DGA 1999) has influenced priorities and roles of other public water agencies and ministries including health, agriculture, environment and electricity (Reyes et al. 2009).

7.5.4 *Managerial Instruments*

Management instruments are sparse. Water rights owners are required to register water rights in the *Registro de Aguas;* the DGA must maintain this information (according to Article 112, Water Code). In the study region the DGA has made a River Depletion Declaration putting a halt to granting new permanent consumptive water rights (Valdez-Pineda et al. 2014).

Juntas de Vigilancia and irrigation associations manage surface water. The National Environmental Commission was tasked with formulating and implementing a National Strategy for Integrated Hydrological Basin Management (NSIBM) in 2009 (Reyes et al. 2009; CNR 2015; 2005), but this process was abandoned after a change in Chilean President.

7.6 Adaptive Governance and Problem Structuring

Figure 7.2 illustrates the instruments responding to d&f in Coquimbo by policy problem quadrant.

The unstructured Q3 problems of **climate change** are addressed by strategies and plans (developed in the capital Santiago far from the study region), with little

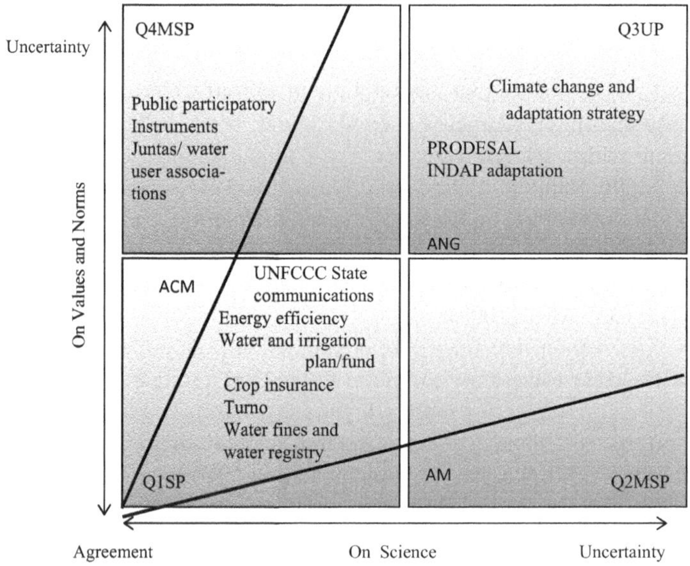

Fig. 7.2 Coquimbo's instruments and problem structuring
ACM Adaptive co-management, *Q4ACM* Quadrant 4 moderately structured problem, *ANG* Anticipatory governance, *Q3UP* Quadrant 3 unstructured problem, *Q2MSP* Quadrant 2 moderately structured problem, *AM* Adaptive management, *Q1SP* Quadrant 1 structured problem

concrete action. Chile frames problems predominantly in Q1, addressing the embedded problem of **drought**. Technological solutions dominate drought response with the building of irrigation and dams to solve issues in an already water challenged context, making drought a structured problem (Q1). This reflects a technocratic management of environmental problems. Similarly **floods** are responded to through structured policies of mapping or building containment infrastructure (in Q1) (Cepeda 2008), albeit these are absent in Coquimbo.

Policies that facilitate **adaptation** are framed in relation to structured Q1 issues of agricultural development, aimed at expanding Chile's export food economy, expanding irrigation, or assisting small agricultural producers (Reyes et al. 2009) without any link to the Q3 problem of climate change. Adaptation focuses on technical expertise in hydrology and economics at the government level and does not account for uncertainty, integration of climate change or inter-annual variability.

Environmental governance approaches are rare as the market manages all water transactions. Public participatory instruments include extensive consultation on the water law amendment in 2014, a Q4 process; however, this consultation occurred in Santiago, far from the study site. The *Juntas* and irrigation associations manage water under their jurisdiction in a Q4 co-management process, but without participation of non-irrigators. Again, no examples of Q2 moderately structured problems (with disagreement on science) were discovered (except in one situation where a disagreement on science could not be resolved and prevented a reservoir project from proceeding (Clarvis and Allan 2013)).

7.7 Impact of Instruments on Actors Measured by Mandate Effectiveness

Regulatory, suasive and managerial instruments are predominantly ineffective in Chile, although there have been some positive experiences with water co-management instruments. Economic instruments have been very effective at creating irrigated agriculture and agriculture exports.

7.7.1 *Regulatory Instruments*

Regulatory instruments have not worked well in Chile. In respect of **climate change**, Chile's extensive plans, strategies, and engagement in programmes haven't shown significant effectiveness as GHG emissions are rising by 57.4% from 1993 to 2006 (UNCCS 2014) and vulnerable populations are at risk of suffering damage due to d&f (Vos et al. 2010). Although as a developing country Chile's emissions may increase, it should try to taper off the rate of growth of its emissions.

In general, regulatory instruments on **water** are not effective and not enforced (Larrain 2014; Donoso 2012). First, there are 43 institutions with overlapping functions, contradictory policies and poor coordination (World Bank 2013). Local governments and local water governance are relatively ineffective (Hurlbert and Diaz 2013). Different sections of a river are managed by independent, non-communicative *Juntas* starved for resources, information, and with little understanding of their mandate (ibid.). Second, there are no mandatory priorities in water use, not even for drinking water (Reyes et al. 2009; Hill 2013; Donoso 2014a, b). Third, the disadvantages of the national water system include strategic water requests by large powerful entities, hoarding of water rights, barriers for development projects, and barriers of entry into regional markets, all discouraging competition and creating monopolies in some cases (without payments to the government) (Brown and Pena 2003). Fourth, many regulatory provisions are not implemented including licensing of groundwater (World Bank 2011) and fines to reduce hoarding (Hill 2013).

Regulatory instruments responding to d&f are ineffective. DOH recognizes that **floods** damage municipal infrastructure (impacting potable water communities, irrigation infrastructure, and roads); however, urban expansion continues in flood risk areas (Reyes et al. 2009). The **drought** declarations and interventions of DGA hinder long-term climate planning (Clarvis and Allan 2013), and heightens conflict because of DGA's low institutional capacity, lack of information on water rights, and low transparency. Technical guidelines in times of *turno* are seen as unclear, confusing, and obsolete and there is a chasm between the current hydro-climate reality and data upon which rights allocation is based (Hill 2013). An expert interviewee concluded that drought declarations prevented businesses from accessing bank credit (Co5), although Hill (2013) concludes that companies use these declarations to access groundwater. The interviewee didn't believe that *turno* was effective, given that an emergency declaration had been made annually for the past 7 or 8 years (Co5); however the extent of the drought was concealed as producers didn't want to disclose to the credit institutions how much their water had been reduced (Co3).

Public participation instruments are not viewed as effective. Environmental impact assessments are viewed by interviewees as having too short a process and no real legal weight as the projects are "already politically pre-approved" (Reyes et al. 2009: 14).

7.7.2 Economic Instruments

Economic water instruments have advanced export agriculture, but the efficiency of the Chilean water market is questioned (Studies supporting efficiency: Rios and Quiroz 1995; Hearne and Easter 1995; Gazmuri and Rosegrant 1996; Gomez-Lobo and Paredes 2001; Cristi and Trapp 2003; Hadjigeorgalis 2004; Studies concluding poor efficiency: Hadjigeorgalis and Iriquelme 2002; and conclusions of variable performance across the country Donoso 2012). Water interests can be hoarded (Bauer 1998). The water market lacks transparency (Donoso 2012). Further, the transfer of water rights faces physical infrastructural barriers (Brown and Pena 2003).

Although some economic instruments stabilize producer income in times of d&f, farmers complain that emergency aid is "too late and too little" (Reyes et al. 2009: 19) and one interviewee felt it had no impact (Co2). The Clean Development Mechanism (CDM) helps Chile attract foreign investment in technology projects; carbon credits are treated as just another export product with little or no contribution to sustainable development (Rindefjall et al. 2011).

7.7.3 Suasive Instruments

Many national and sectoral adaptation plans have been formulated but these plans have very little application in local concrete actions, especially in the study region. In Santiago one adaptation plan exists (Krellenberg et al. 2012) and one adaptation project is underway (Amur 2013).

7.7.4 Managerial Instruments

Somewhat contradicting the water market system, and confirming customary practice, historical hydrological conditions, agreements, and rules govern reservoir operation (Vicuna and Meza 2012). Informal water practices persist despite the

1981 Water Code (Boelens 2013), but nationally the Chilean water market is regarded as the predominant institution.

Informal agreements of cross-sector collaboration between agricultural stakeholders, mining companies, and *Juntas* govern water quality in some watersheds. Some view these as 'payoffs' for mining impacts on water quality; others as corporate social responsibility (Hill 2013: 206). Interviewees felt that these public private water tables[4] were a good experience. However, the rigid institutional structure surrounding water affects initiatives to integrate basin water management for improving the environment (Reyes et al. 2009). Basin organizations and water tables are consistently underfunded, undertaking short-term initiatives with little impact (Dononso 2014b). The planned water registry hasn't been implemented; this allows the water market to continue to operate without transparency as a 'dark market' (Hill 2013: 143–148).

Currently, inter-municipal regulatory plans, local regulatory plans, local development plans, sectional plans, regional development strategies and plans are fragmented, underutilized for adaptation, and do not take into account climate change (Barton 2013).

7.8 Assessment of Learning

Table 7.3 assesses how the instruments contribute to social learning. The most significant social learning is double loop learning where new strategies changed underpinning assumptions. This occurred when irrigation at all levels was expanded by the development of a water market, irrigated agriculture, and an export agricultural economy capable of doubling GDP in the past 10 years (Valdez-Pineda et al. 2014). The 1981 Water Code efficiently increases investment due to the security of water rights granted in legislation (Donoso 2008, 2012) raising scholarly interest (Bauer 2004; Domper-Rodriguez 2009; Pena 2006). However, the continuation of this paradigm, given climate change, degraded ecosystem services, and growing inequality, risks zero loop learning. Zero loop learning might also result with stranded irrigation infrastructure.

Single loop learning has occurred in rural dry land communities where people have adjusted to the reality of water scarcity (Cortes 2010) although no instrument can be identified as responsible. Farmers have a deep knowledge base, sensitivity and experience in responding to drought periods, despite lack of formal training. They know when to restrict their use and even when it will be a hydrologically scarce growing season. It is said to "be in their DNA" (Hill 2013: 208).

Dam projects in Coquimbo display zero loop learning. Five communities were relocated when the Pulcaro dam was built in 1995, one of which ended up with no access to water, farm land or even small gardens (Rojos et al. 2007). Further up the

[4] Water tables are a formal institution where all parties with an interest in water meet to discuss issues, exchange knowledge, and potentially come to agreement on water issues.

Table 7.3 D&F: Assessment of learning in Coquimbo

Level/learning evidenced	Change in practice – single loop learning (one way information flow – Q1)	Change in assumptions or models – double loop learning (two way information flow Q2 Q4)	Change in values, norms, world views or power dynamics – triple loop learning (iterative information flow – Q4)
Individual	Changing agricultural practices, e.g. crops, planning times	Expansion of irrigation	–
Regional	Changing agricultural practices	Expansion of irrigation	–
National	Changing agricultural practices	Expansion of irrigation	–

Andes, the Canadian mining company of Barrick Gold and Chilean and Argentinian governments ignored public opposition rendering their participation in environmental reviews 'superficial' (*ibid*: 64). Further, local water groups have little power and their effectiveness at contributing to water and development issues in Q3 or Q4 appear limited. In relation to disasters there is a lack of learning (Co4).

One interviewee stated that people including water users (producers) and regional organizations, wanted to participate in public policy, but the government had little interest in listening (Co2). Two issues present opportunities for learning: (1) the private system of water interests and modifying it to recognize the human right to water (Donoso 2014b); and (2) adaptation as a result of the successive declarations of emergency. Within these discussions CSOs demand the human right to water and local water committees assert presence in relation to the wicked problem of adaptation. Overall Q3 is vacant; occasional dialogues such as these have yet to be productive and are distant from Coquimbo.

7.9 Instrument Impacts on Livelihood Capitals

Table 7.4 summarizes how the instruments impact agricultural producer livelihood capitals. Instruments have benefited large agricultural producers' economic capital. These producers then have access to information (human capital), efficient irrigation and adaptation measures (technological capital), and water rights (natural capital). Small and medium sized producer livelihoods are not as resilient.

Table 7.4 Assessment of main instruments

Type	Instrument	Eff	+ Impact on Capitals	(−) Impact on Capitals
Reg	Fee for non-use of water right	–	–	– Natural
	Holdback for minimum river flow	–		– Natural
	Fine for illegal water extraction	–		– Natural
	Presidential Declaration of Drought -*Turno*	–		– Economic
But miss	Enforceable conditions of water property interest to recognize social nature of water			
	Emergency measures planning requirements			
	Building codes			
	Improved and integrated groundwater governance			
	GHG mitigation			
Econ	Full water property interest allows water market	+	+ Economic	
	Irrigation funding	+	+ Economic	–
	Small and medium producers 75% of investment	+	+ Economic	
	Irrigation or efficiency upgrade		+ Technological	
	Loan instruments	+	+ Economic	
	Direct government subsidy of drinking water for low income families	+	+ Human	
But miss	Debtor forgiveness in bankruptcy			
Suas	National Plan on Climate Change Adaptation (eight sector plans in development (one is water))	–	–	–Natural
	Glacier Protection Policy	–		–Natural
	National Climate Change Action Plan	–		–Natural
But miss	Agriculture environmental incentives			
	Public information on water, groundwater, climate			
	Public Water Rights (drinking, tourism, landscape preservation, sport fishing, rccrcation)			
Mgt	National irrigation plan	–	+ Economic	–
	Management of canals, reservoirs or aquifers by Jd V or water groups	+	+ Social	
But miss	Demand management of water			
	National disaster mitigation strategy for d&f			
	Drought policy and drought research and planning program			
	Long term water management plans on integrated basis			
	Source water protection plans			
	Inclusive, participatory development instruments			

Inst Instrument, *Reg* Regulatory, *Miss* Missing, *Econ* Economic, *Suas* Suasive, *Mgt* Managerial
++ Effective; + moderately effective; – Ineffective

7.9.1 *Human Capital*

The existing poverty, poor health and education and d&f are exacerbated by trade liberalization and government austerity. Within the fifteen municipalities in Coquimbo Region, the average poverty rate is 25% (Kubik et al. 2010). In rural areas, approximately 4.5% of the total population (some 700,000) don't have access to drinking water and there is little waste treatment (Brown and Pena 2003: 18). This worsens during drought and flood (Larrain 2014). Instruments of water quality aren't enforced due to lack of resources (Reyes et al. 2009).

Deficits in education exist. The younger generation increasingly migrates to other areas to find employment (Vergara and Barton 2012), dropping out of high school, and thereby impeding poverty reduction measures (Reyes et al. 2009). There is a lack of information and education on climate, climate change, water management, water conservation, and drought preparation (Hill 2013). There is a lack of professional training and education in water resource management (Hill 2013; Kubic et al. 2010; Brown and Pena 2003).

7.9.2 *Social Capital*

There is a high level of social stratification. Large producers, generally Italian immigrants from the 1950s, produce grapes for export, domestic consumption, and *pisco*. Some are multinational companies such as El Monte or Dole who produce fruit or vegetables for national or international sale. Medium scale farmers own over 12 irrigated hectares for producing grapes, vegetables and fruit, utilize family members for labour, and depend on state loans of the INDAP (Salas et al. 2012). Small farmers (including goat breeders) depend on labour income in the agro-industrial farms or mines and grow produce for local sale or family consumption (Salas et al. 2012).

This stratification reduces social capital and access to technological capital for small and medium producers (see Sect. 7.9.4). They have less access to information and resources to conduct technical analysis to maintain and defend water rights and determine the value of rights (Brown and Pena 2003). There is little trust in the water governance system resolving conflicts, as the state is not seen as a neutral arbitrator as it owns and controls a national mining company with low accountability (Reyes 2014)). Private rights holders effectively manage water and the strength of their participation is dependent on the amount of rights owned.

The *Juntas* generally lack social cohesion; there may or may not be a *Junta*; it may or may not function properly or be legalised; there may or may not be cooperation and trust between members; and often there are rivalries, practices of unproportional distribution of water, power imbalances, and a lack of cooperation (Hill 2013; Donoso 2014a). Barring some exceptions, most water user associations do not manage the water resource in accordance with water laws (Donoso 2014a, b).

Because of the high autonomy of individual water rights owners, the *Junta* often doesn't have a sufficient mandate to be effective, and public authorities lack agency and capacity to invoke water resource policies and practices to encourage cooperation across a watershed and ensure sustainability.

7.9.3 Economic Capital

There is a widening gap between large capital intensive viticulture producers and small and medium ones (OECD 2008). Large operations can invest in water rights, land, modern equipment, hire well-trained agronomists and wine makers, and take advantage of the shifts in optimum growing conditions (Hadarits et al. 2016). Those with more resources can purchase additional water rights (Valdez-Pineda et al. 2014) and assert these rights in the formal water market and court system to the disadvantage of those relying on traditional water practices.

The initial water allocations were not, and still are not, transparent. Negotiation between government officials, members of Parliament, and companies occur with agreements resulting in exemptions from licence fee payments in the millions (CDCA n.d.). This transparency is further strained due to lack of and differential access to information. Information is "a sort of 'elitist informational condition'" (Salas et al. 2012: 39) whereby some stakeholders lack access to information either because it is not made public or because it circulates through internet channels that not everyone has access to.

During drought and water shortages (2009–2014) municipalities pay for trucking drinking water to residents while large producers and mining companies continue production (Larrain 2014). Free water rights were granted by the state to mining companies, irrigators, and industrial users; but now in order to provide drinking water, the state has to compensate for the expropriation of these rights (Larrain 2014). The water law works against small landholders, many of whom have no water rights and are dependent on the agricultural industry. Only half the small-scale farmers can access credit and subsidies provided by INDAP (OECD 2008).

There are fewer livelihood opportunities and employment is typically seasonal with low wages (Young et al. 2010). In the dry land areas such as the study region, drought decimates livestock herds and small agricultural production. This creates a chain of social problems – low revenue, lack of food, thereby necessitating migration to urban centers for paid work (Salas et al. 2012).

7.9.4 Technological Capital

In Chile, investments in large water reservoirs and other large investments in the Coquimbo region have created technological capital contributing to water security for agriculture (Reyes et al. 2009: 54). This strength overdone is also a weakness. Technology is regarded as the cure for all ills and inequitably dispersed.

Large, capital intensive viticulture operations, especially those growing grapes and producing wine, are flexible and can adopt strategies to reduce drought risk by making changes to their vineyard and their winery, adopting different grape varieties, different technologies, and purchasing land in different locations (Hadarits et al. 2016). Large solvent producers can access the necessary economic instruments to exploit additional groundwater resources during drought, when this is allowed as a short-term strategy (Clarvis and Allan 2013). Small farmers and farming communities can scarcely access irrigation as they lack or have unclear water rights and have unclear land titles (Reyes et al. 2009).

Because of the historical focus on economic parameters, innovation is generally low, relying on classical fixes such as large-scale dam storage and increased groundwater exploitation (Clarvis and Allan 2013). Programmes tend to expand irrigation, rather than save water (*ibid.*). In small communities, irrigation channels are only made of soil, which results in water loss by evaporation and infiltration and financial resources can't be accessed to correct this (Kubik et al. 2010).

7.9.5 Natural Capital

Ecosystems, rivers and aquifers have become compromised (Donoso 2012). As legal extraction can exceed water availability, water rights are effectively only "paper rights" (Hill 2013: 230; Donoso 2012; Pena 1997, 2006). There is little incentive for farmers to reduce their water use, or sell it, as it will be needed when there is a proportional reduction in the basin in dry years (Hill 2013). Water rights holders with paper rights also hold on to these in anticipation of a particularly wet year when their rights might be fulfilled. **Drought**s are exacerbated by illegal extractions (which can't be curtailed as legal recourse is currently the only tool). The environment is deteriorating because of reliance on exploited, unregulated, groundwater by large producers, industry, and local municipal governments (ibid.; Co1). Industrial development increases salt deposits and reduces wetlands (Larrain 2014). The increasing urban water demand, planned increases in irrigated agriculture, and projected reductions in water available from glacier melt (glaciers are not protected by law), and reduced precipitation and temperature rise, does not bode well for future natural water capital (Barton 2013).

No regional network of institutions monitors water quality and quantity. Instead, private companies are outsourced for ad hoc studies; there is no consistency and agreement on the status of water resources, their health, and how much illegal water usage is occurring, etc. (Hill 2013). Water pricing does not differentiate between use or sector. Hence, there is no incentive for demand side or water balance management (Hill 2013).

7.10 Re-designing Instruments with the ACW

This chapter explored how Coquimbo's institutional policy framework impacts capitals of agricultural producers. Based on this, the adaptive capacity of Coquimbo is assessed in Fig. 7.3 and recommendations made to enhance adaptive governance (see Table 7.5).

Chile's ACW includes both high and low ratings. Coquimbo received a low rating in relation to **variety** of problem frames, instruments, and solutions. Technology (Q1 water diversions) and the market system are prioritized and this reduces Coquimbo's rating in relation to redundancy and diversity. The federal government and private market predominate, reducing the number of actors. There is a deficit in instruments responding to climate change and flood, or tackling issues where there is disagreement on science or on values and norms.

Gains in resources and leadership have occurred because of the double loop **learning** resulting from the adoption of irrigation instruments together with the

Green	Light Green	Yellow	Light Orange	Red
Very High	High	Medium	Low	Very Low

Fig. 7.3 Chile's ACW

Table 7.5 Redesign options for Chile

Analysis unit	Description
Ineffective instruments	Regulatory – ecological flow, water fees, fine for non use, river holdback, water registry, local water management by juntas, local water groups
	Presidential declaration of drought – *Turno*
	INDAP emergency funds
	Climate change adaptation plans, Glacier Protection Policy, water plans, flood zone infrastructure plans
Missing instruments	Climate change mitigation and adaptation, carbon market, GHG reductions
	Enforceable water property interest to recognize social nature of water, human and public right water, integrated groundwater regulation, public information surrounding water, demand management of water
	Emergency planning requirements, building codes, flood zone restrictions, national disaster mitigation strategy for d&f, drought policy, and drought research and planning
	Insolvency
	Agriculture environmental incentives
	Inclusive participatory development instrument
Unaddressed drivers	Climate change and ecosystem deterioration
	Neo-liberal market and colonial elitism
	Increasing trade liberalization, input costs, and energy demand
	Urbanization, population growth, aging producers, shortage of farm labour
	Growing inequality
	Conflict of informal water practices with legal rules
Weak adaptive capacity dimension	Fair governance, equity
	Variety (diversity)
	Learning capacity (trust)
Constrained livelihood capital	Human (discrepancies of poor)
	Social
	Economic of small producers
	Natural capital (drought – 7 years of emergency declaration)

private property interest in water. This and the resilience of small producers facilitate single loop learning in new techniques. However, the neo-liberal market model predominates and no change in assumption or models inherent in triple loop learning has emerged. The double loop learning is in jeopardy as there is no learning in relation to extreme events and the past 7 years of emergency drought declarations are now a 'new normal.' Having such a significant drought instrument is significant, but it has not been recalibrated utilizing principles of adaptive management or anticipatory governance.

Trust is often not present; doubts are discussed frequently with no resolution, and the market and judicial system is accused of 'nepotism' (Hill 2013). Although water privatization has assisted the initial double loop learning of irrigation, given climate change and recurring drought, it has prevented many people from realizing their human right to water (Reyes et al. 2009), further eroding equity and trust. In combination with government austerity and neoliberalism, this privatization has been detrimental for small and medium producers. Growing inequality and deteriorating ecological services threaten the historical double loop learning of irrigation as more Chileans live in poverty without access to drinking water. Local governments without resources attempt to meet this need.

There is much room for **autonomous change** and **leadership** by powerful interests with water rights. These groups have **resources** (Hadarits et al. 2016), but small and medium producers do not. Targeted programmes for small producers are often unattainable because of their lack of water rights and they are further excluded from domestic and international markets through continued practices of hegemony by the state and large agricultural producers reinforced by neoliberalism and colonial elitism. **Human and economic resources** on the ACW are ranked yellow, or neutral as they are very high for large producers, and very low for small. **Fair governance** and **equity** are a problem. **Collaborative** leadership is strong at the local community level with water groups (Reyes et al. 2009) resolving problems without assistance from other levels of government.

The most important redesign option for Coquimbo to improve adaptive governance involves addressing growing inequality (see Table 7.5). Trade liberalization, government austerity, and colonial elitism have combined with the private water market to favour large agricultural producers and industries and erect insurmountable barriers for small and medium producers in relation to access to water, technology, health, economic, and social capital, and participation in development decisions. Deteriorating natural capital, the reduction of water evidenced by the past drought declarations (seven in total), and the hesitance of agricultural producers to report their reduced water allocations to their financial institutions is concerning.

Current regulatory instruments need to be strengthened with government resources (financial and human) and management and suasive instruments need to have concrete implementation instruments. Missing instruments including the human right to water and sanitation services, local water governance, integrated surface, and groundwater instruments. Inclusive participatory development instruments are also missing; this will allow the voices of small, marginalized producers to be heard and contribute to resolution of policy problems as well as normalizing local water practices (Boelens 2013).

Unaddressed drivers of the neo-liberal market, colonial elitism, and demographic changes require instruments calibrated to improve the constrained livelihood capitals of human health, education, and access to information. Social capital issues experienced by small and medium producers require attention. Although there is strong bonding social capital to one another within local agricultural and water groups, the linking social capital to people in authority is absent. The requirement of interconnection between these local groups and the organizations and institutions of the Chilean government located in Santiago, as well as large powerful state and non state actors is evident in this case study, and therefore a necessary component of adaptive governance.

This case study also illustrates the importance of assessing and reassign learning within adaptive governance over time (a finding that also emerged in Alberta). Although the building of dams and funding of irrigation has advanced learning in the past, the same dams have negatively impacted affected communities and small agricultural producers without water rights. Including small and medium producers in authentic participatory development planning can target the adaptive capacity dimensions of fair governance (equity), trust, and prevent projected zero loop learning (stranded infrastructure and exacerbated inequality) in relation to irrigated agriculture.

References

Amur (2013) Mapa: un plan de adaptacion a la variabilidad y cambio climatico para la Cuenca del Rio Maipo. Newsletter, June 2013. Available at: http://www.amur.cl/web/mapa-un-plan-de-adaptacion-a-la-variabilidad-y-cambio-climatico-para-la-cuenca-del-rio-maipo. Accessed 12 Dec 2014

Barton J (2013) Climate change adaptive capacity in Santiago de Chile: creating a governance regime for sustainability planning. Int J Urban Reg Res 37(6):1916–1933

Bauer C (1997) Bringing global markets down to earth: the political economy of water rights in Chil. 1976–95. World Dev 25(5):639–656

Bauer C (1998) Against the current: privatization, water markets, and the state in Chile. Kluwer Academy, Norwell

Bauer C (2004) Siren song: Chilean water law as a model for international reform. Resources for the Future, Washington, DC

Boelens R (2013) Cultural politics and the hydrosocial cycle: water, power and identity in the Andean highlands. Geoforum. http://10.1016/j.geoforum.2013.02.008. Accessed 12 Dec 2014

Brown E, Pena H (2003) Systemic study of water management regimes, Chile. December, 2003. Global Water Partnership, South America

Budds J (2004) Power, nature and neoliberalism: the political ecology of water in Chile. Singap J Trop Geogr 25(3):322–342

Carruthers D (2001) Environmental politics in Chile: legacies of dictatorship and democracy. Third World Z. 22(3):343–358

CDCA (Centro Documentazione Conflitti Ambientali) (n.d.) Dams on the Baker and Pascua Rivers – Endesa/Enel Chile. Available at: www.cdca.it/spip.php?article1646&lang=en. Accessed 12 Dec 2014

Cepeda J (ed) (2008) Los Sistemas Naturales de la Cuenca del Rio Elqui (Region de Coquimbo, Chile): Vulnerabilidady Cambio de Clima. Universidad de La Serena, La Serena

Clarvis MH, Allan A (2013) Adaptive capacity in a Chilean context: a questionable model for Latin America. Environ Sci Policy. Available at: http://dv.org10.1016/j.envxxc.2013.10.014. Accessed 12 Dec 2014

CNR (2005) Politica Nacional de Riego, approved by Council of Minister in Dec 2005 in www.cnr.cl

CNR (2015) Comision Nacional de Riego. Available at: www.cnr.cl. Accessed 16 April 2015

CONAMA (2006) Estrategia Nacional de Cambio Climatico. Government of Chile, Santiago

CONAMA (2008) Plan de Accion Nacional de Cambio Climatico. 2008–2012. Government of Chile, Santiago

Cortes EJ (2010) Reflections of the La Serena Participants – participatory action research in the semi-arid North of Chile. In: Kubik W, Corkal D, Rojas A, Sauchyn D (eds) Rural communities adapting to climate-induced water stress. Canadian Plains Research Cente, Regina

Cristi O, Trapp A (2003) Mercado de agua para Irrigacion y Uso Urbano: una Aplicacion a la Cuenca del Rio Elqui, Chile. World Bank-Netherlands Water Partnership Programmes (BNWPP), Water Rights System Window, July 2003

DGA (Direccion General de Aguas) (2007) Fuente: Balance de Gestion Interal. Direccion General De Aguas. Available at: www.dipres.cl

DGA (1999) Política Nacional de Recursos Hídricos. Available at: http://www.cepis.ops-oms.org/bvsarg/e/fulltext/chile2/chile2.pdf. Accessed 12 Dec 2014

de la Domper-Rodriguez ML (2009) Chile: a dynamic water market. Libertad y Desarrollo: an economy think tank, pp 1–9. Available at: http://www.fcpp.org/pdf/09-03-23-Chile.pdf. Accessed 12 Dec 2014

Donoso G (2008) Water use rights markets and water allocation: the Chilean Case Expo Zaragoza 2008. Thematic week: water economics and financing thematic axis. Water Markets. http://www.zaragoza.ex/contenidos/medioambiente/cajaAzul/24S7-P3-Guillermo%20DonosoACC.pdf. Accessed 12 Dec 2014

Donoso G (2012) The evolution of water markets in Chile. In: Maetsu J (ed) Water trading and global water scarcity: international perspectives. RFF Press, Washington, DC

Donoso G (2014a) Presentation to workshop. From global to local: sustainable water resource governance for Latin America and the Caribbean. Universidad Catolica de Chile, Santiago Chile, May 5, 2014

Donoso G (2014b) Presentation of "Los desafios de la institucionalidad del agua en Chile" to Seminario Internacional, Governanza del Aqua: Desde lo global a lo local, Universidad Catolica de Chile, Santiago Chile, May 7, 2014

Dourojeanni A (2014) Presentation as Fundacion Chile "La gobernanza del aqua a nivel latinoamericano" to Seminario Internacional, Governanza del Aqua: Desde lo global a lo local, Universidad Catolica de Chile, Santiago Chile, May 7, 2014

Ernst & Young (2013) In: London UK (ed) Latin America insurance outlook. Posed for Growth, EYGM Limited

FAO (Food and Agriculture Organization) (2003a) United Nations FAO roles of agriculture Chilean case study reports. ftp://ftp.fao.org/es/ESA/roa/pdf/ [links to Chilean case studies] Perceptions of environmental role of Chilean Agriculture. Available at: ftp://ftp.fao.org/es/ESA/roa/pdf/2_Environment/Environment_Chile2.pdf Accessed 13 April 2008

FAO Food and Agriculture Organization (2003b) United Nations FAO roles of agriculture Chilean case study reports. Available at: ftp://ftp.fao.org/es/ESA/roa/pdf/ [links to Chilean case studies] Synthesis Report: ftp://ftp.fao.org/es/ESA/roa/pdf/NR/NR_Chile.pdf. Accessed 13 April 2008

Gazmuri R, Rosegrant M (1996) Chilean water policy: the role of water rights, institutions, and markets. Resour Develop 12:33–48

Gomez-Lobo A, Paredes R (2001) Reflexiones sobre el proyecto de modificacion del Codigo de Aguas. Estudios Publicos 82:83–91

Gobernanza del agua (2014) Presentation to workshop, from global to local: sustainable water resource governance for Latin America and the Caribbean at Universidad Catolica de Chile, Santiago Chile, May 5, 2014

GC (2010) National climate change action plan. Santiago, Chile. Available at: www.mma.gob.cl/1304/articles-49744_Plan_02.pdf. Accessed 12 Dec 2014

Hadarits M, Santibanez P, Pittman J (2016) Chapter 13: Drought risks and opportunities in the Chilean grape and wine industry: a case study of the Maule region. In: Diaz H, Warren J, Hurlbert M (eds) Vulnerability and adaptation to drought: the Canadian Prairies and South America, ISSN 1919-7144. University of Calgary Press, Calgary, pp 303–326

Hadjigeorgalis E, Iriquelme C (2002) Analisis de los Precios de los Derechos de Aropvechamiento de Aguas en el Rio Cachapoal. Ciencia e Investigacion Agraria 29(2):91–99

Hadjigeorgalis E (2004) Comerciando con Incertidumbre: Los Mercados de Agua en la Agricultura Chilena. Cuadernos de Economia 40(122):3–34

Hannah L, Roehrdanz PR, Ikegami M, Shepart AV, Shaw R, Tabor G, Zhi L, Marquet PA, Hijmans RJ (2013) Climate change, wine and conservation. Proc Natl Acad Sci 110:6907–6912

Hearne R, Easter W (1995) Water allocation and water markets: an analysis of gains from trade in Chile, WBTP 315. World Bank, Washington, DC
Hill M (2013) Climate change and water governance. Adaptive capacity in Chile and Switzerland. Springer, Dordrecht/Heidelberg/New York/London
Hurlbert M, Diaz H (2013) Water governance in Chile and Canda- a comparison of adaptive characteristics. Ecol Soc 18(4):61–76
INDAP (2015) Local development programme (PRODESAL). Available at: indap.gob.cl/programas;programa-de-desarrollo-local-prodesal
Krellenberg K, Barth K, Barton J, Harris J, Hansjurgens B (2012) Plan de Adaptacion al cambio climatico para la Region Metropolitana de Santiago de Chile. STG, CAS. Ministre del Mefdio Ambiente, Santiago
Kubik W, Corkal D, Rojas A, Sauchyn D (eds) (2010) Rural communities adapting to climate-induced water stress. Canadian Plains Research Center, Regina
Larrain S (2014) Presentation as Directora Progama Chile Sustentable to Seminario Internacional, Governanza del Aqua: Desde lo global a lo local, Universidad Catolica de Chile, Santiago Chile, May 7, 2014
MAPS (2015) Mitigation action plans and scenarios. Available at: http://www.mapsprogramme.org/projects/chile-projects/maps-chile-overview. Accessed 18 Apr 2015
McCord MJ, Ingram M, Tatin-Jaleran C (2013) The landscape of microinsurance in Latin America and the Carribean. Available at: http://www.fomin-events.com/foromic/2013/_upload/panelistas/36_3Njy5.pdf. Accessed 19 Apr 2015
Nachmany M, Fankhauser S, Townshend T, Collins M, Landesman T, Matthews A, Pavese C, Rietig K, Schlieifer P, Setzer J (2014) Chile extract from The GLOBE climate legislation study, a review of climate change legisltion in 66 countries. (4th Edition). Globe International 2014, UK. Available at: http://www.lse.ac.uk/GranthamInstitute/publications/Policy/docs/Globe2014.pdf. Accessed 12 Dec 2014
ODEPA. (Oficina de Estudios Y Politicas Agrarias) (2013) Estadisticas [online] ODEPA. Available: www.odepa.gob.cl/menu/MacroRubros.action;jsessionid=0915FC15FFEDD893CF88D3A33224DBED?rubro=agricola&reporte=. Accessed 12 Dec 2014
OECD (2008) OECD review of agricultural policies: Chile 2008. OECD Publishing
OECD (2011) Water governance in OECD countries: a multilevel aproach. OECD Studies on Water, OECD Publishing. Available at: http://dx/doi.org/10.1787/9789264119284-en. Accessed 12 Dec 2014
OECD (2013) Water and climate adaptation; policies to navigate uncharted waters, OECD, studies on water. OECD Publishing. Available at: http://dx.doi.org/qp.1787/9789264200449-en Country Profile Chile. Accessed 12 Dec 2014
Pena H (2006) Chile's experience on water reform. Ministry of Public Works, Chile, pp 25–26
Pena H (1997) Modificaciones al Codigo de Aguas y su Aporte a la Gestion del Agua. Direccion General de Aguas, Ministerio de Obras Publicas, Santiao de Chile
Peralta F (2014) Presentation as Presidente de la Confederacion de Canalistas to Seminario Internacional, Governanza del Aqua: Desde lo global a lo local, Universidad Catolica de Chile, Santiago Chile, May 7, 2014
Pizarro JC (2009) Institutional adaptations to climate change: comparative study of dryland river basins in Canada an d Chile. University of Regina (Canada) and Universidad de La Serena (Chile) Working Paper in the IACC. dated March 2, 2009
Reyes B, Salas S, Schwartz E, Espinoza E (2009) Chile governance assessment final report. Available at http://www.parc.ca/mcri/pdfs/papers/gov03.pdf Accessed 12 Dec 2014
Reyes F (2014) Presentation to workshop. From global to local: sustainable water resource governance for Latin America and the Caribbean. Universidad Catolica de Chile, Santiago Chile, May 5, 2014.
Rindefjall R, Lund E, Stripple J (2011) Wine, fruit, and emission reductions: the CDM as development strategy in Chile. Int Environ Agree 11:7–22
Rios M, Quiroz J (1995) The market of water right in Chile: major issues. World bank technical paper 285. The World Bank, Washington, DC

Rojos Q, Reyes B, Magzul L, Morales HL, Borquez R, Swrtz E (2007) Analysis of the role of institutions in water conflicts final report. IACC project report. Available at: http://www.parc.ca/mcri/iacc057.php. Accessed 12 Dec 2014
Romero H, Mendez M, Smith P (2012) Mining development and environmental injustice in the Atacama Desert of Northern Chile. Environ Justice 5(2):70–77. doi:10.1089eenv.2011.0017
Salas S Jimenez E, Montaña E, Garay-Fluhmann RG, Gauthier D, Diaz HP (2012) Vulnerability to climate change. Challenges for adaptation in the Elqui and Mendoza basins. Inter-American Institute for Global Change Research, La Sarena
SISS. (Superintendencia de Servicios Sanitarios) (2015) Webpate of Gobierno de Chile. Available at: www.sess.gob.cl. Accessed 12 Dec 2014
UNCCD (2013) Bakcground document. The economics of desertification, land degradation and drought: methodologies and analysis for decision-making. Pre-pared for 2nd scientific conference, April 9–12, 2013 Bonn, Germany
UNCCS (2014) Available at: https://unfccc.int/files/ghg_data/ghg_data_unfccc/ghg_profiles/application/chl_ghg_profile.pdf. Accessed 16 Sep 2014
UNFCCC (2015) Homepage. Available at: unfccc.int. Accessed 30 Mar 2015
Valdez-Pineda R, Pizarro R, Garcia-Chevesich P, Valdes JB, Olivares C, Vera M, Balocchi F, Perez F, Vallejos C, Fuentes R, Abarza A Helwig B (2014) Water governance in Chile: availabiity, management and climate change. J Hydrol. Available at: http://dx/doi.org/10.1016/j.hydrol.2014.04.016. Accessed 15 Dec 2014
Vergara EP, Barton JR (2012) Poverty and dependency in indigenous rural livelihoods: Mapuche experiences in the Andean foothills of Chile. J Agrarian Change 13(2):234–262
Vicuna S, Meza FJ (2012) Los nuevos desafios para la gestion de los recursos hidricos en Chile en el marco del Cambio Global. Temas de la Agenda Publica. Centro de Politicas Publicas UC. Ano 7, No 55 November 2012
Vos F, Rodriguez J, Below R, Guha-Sapir D (2010) Annual disaster statistical review 2009: the numbers and trends. CRED, Brussels
World Bank (2009) The World Bank, BINI Index. http://data.worldbank.org/indicator/SI.POV.GINI?
World Bank (2011) Chile. Diagnostico de la Gestion de los Recurosos Hidricos. Departamento de Medio Ambiente y Desarrollo Sostenible, Region para America Latina y el Caribe. Chile, 88 pp. http://water.worldbank.org/node/83999> University of Calgary Press
World Bank (2013) Overcoming institutional and governance challenges in environmental management. Case studies from Latin America. Enviornment & water resources occaional paper series. 76895. Washington. International Bank for Reconstruction and Developmenté éThe World Bank
Young G, Cepeda J, Humberto A, Wandel J, Smit B, Salas S (2010) Vulnerability and adaptation in a dryland community of the Elqui Valley, Chile. Climatic Chang 98(1):245–276

Laws

The Water Code, 1981
Law 18,450 of 1985 , the Promotion of Private Investment in Irrigation and Drainage Wroks
Article 19.24 1980Constitution of the Republic of Chile
Law 20.444 Creation of the National Reconstruction Fund. Available at: http://www.preventionweb.net/english/professional/policies/v.php?id=21231 accessed March 31, 2015

Chapter 8
Case Study Mendoza, Argentina

8.1 Introduction

This chapter explores how the capitals of agricultural producers in Mendoza, Argentina are impacted by the institutional policy framework surrounding d&f. The chapter is structured like other case study chapters exploring and evaluating adaptive governance.

Mendoza has experienced 4 years (2010–2014) of emergency drought declarations and implementation of the instrument of water rationing through *turno*. This chapter shows that the most significant unaddressed driver impacting agricultural producers is the Argentinian monetary policy combined with trade liberalization, which creates risk surrounding inflation, cost of credit, and market prices. However, strong social capital combined with a water instrument unique to Mendoza (inherence – see Sect. 8.5.1) and targeted economic instruments for small producers, protect small and medium irrigated producers from loss of livelihood. The same suite of instruments (especially the regulatory inherence water instrument) prevents adaptation as water is proportionally reduced by the process of *turno* without consideration of demand requirements of producers.

Unique circumstances relating to social capital together with strong management instruments for water rights holders, leadership, and iterative public consultations have achieved triple loop learning and protection for glaciers. Although technocratic Q1 policies and responses predominate, Mendoza has had significant consultation over the years in relation to wicked Q3 issues (integrated water management, integrated land planning etc.). A significant cohort of environmentalists, civil society groups and politicians participate in each of these activities and their unique oasis culture and climate change are important issues to them. A strong water management system situated in Q4 created and sustains the Mendoza oasis. When redesigning instruments to address drivers and implementing missing instruments, Mendoza must retain these instruments as these unique instruments create a strong base of adaptive governance.

M.A. Hurlbert, *Adaptive Governance of Disaster*, Water Governance - Concepts, Methods, and Practice, DOI 10.1007/978-3-319-57801-9_8

8.2 The Case Study Area

Mendoza, the study region in Argentina, is located in the southern central Andes Region on the eastern side of the South American Mountains. The climate is Mediterranean. During cold winters snow accumulates on the mountains; during the summer the yearly precipitation occurs in minute amounts. During the spring the winter snow accumulation melts and a rise of river runoff (the Mendoza river) occurs peaking in December and January. This annual runoff provides water to the population for agriculture and hydroelectricity (Montaña and Boninsegna 2016). Extreme drought has been experienced in Mendoza in 2010–2014 (General Department of Irrigation (DGI) 2015) and 1966–1970 (Prieto et al. 2010); floods can be experienced during periods of very intense rains of short duration. These may produce landslides in urban areas and excess moisture is channelled into the irrigation network of the Mendoza River surpassing the transport capacity of the network, flooding property and damaging infrastructure (Diaz and Bertranou 2004).

Intensive, diversified irrigated agriculture consisting of viticulture and horticulture occurs in 3.6% of the Mendoza region (Mussetta 2013). 98.5% of the population live in the irrigated area. A web of medium and small sized towns spread over these agricultural lands away from the capital city of Mendoza (Montaña and Boninsegna 2016). Grape producers with traditional vineyards, viticulture, and horticulturalists (often of Bolivian origin) comprise the irrigators. Goat breeders (*Campesinos* or '*guarpes*') populate non-irrigated lands (the 'desert') (Montaña and Boninsegna 2016) (Fig. 8.1).

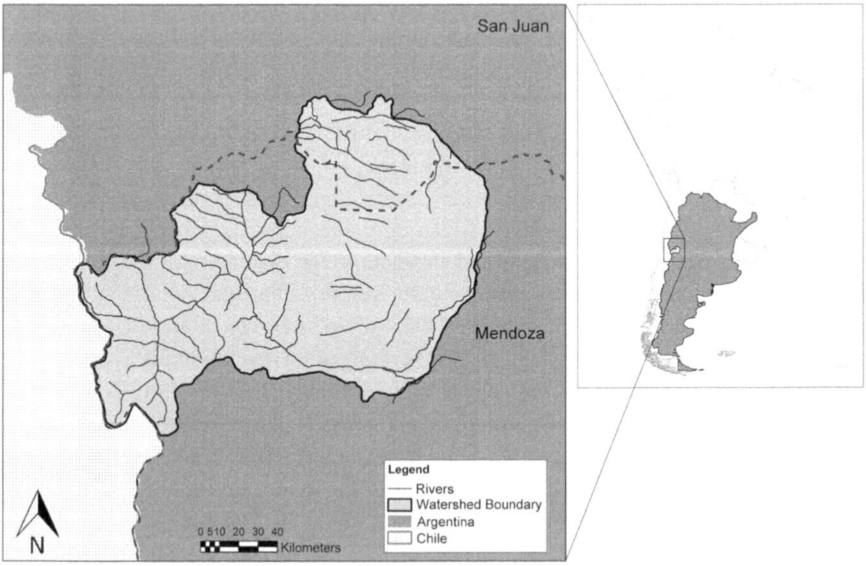

Fig. 8.1 Study region in Mendoza, Argentina

8.3 Main Drivers Impacting Rural Agricultural Producers

Climate change, international economic drivers and Argentina's unique financial position, create a context very different from the other case studies.

Climate change projections indicate a probable decrease of snow in the mountains and a rise in temperature during the present century; this will increase the water deficit and compromise Mendoza's oasis survival (Montaña and Boninsegna 2016). Long-term climate change predictions are for an increase in mean annual temperature of between 2.5 and 3 °C and a reduction in snowfall and runoff of between 10 and 15% per annum. An increase in summer precipitation of about 30% is predicted (*ibid.*; Boninsegna and Villalba 2006). Ecosystem services are deteriorating (M1).

Similar **demographic** drivers exist as in the other case studies. Young rural dwellers are migrating to urban centers and the remaining rural population is aging (Montaña and Boninsegna 2016). Aging farmers are generally resistant to change (M1, M6). The expansion of mining (oil and gas) has created jobs, but is exhausting already scarce water resources and competes for labour with agriculture (Montaña et al. 2005). This driver has unique implications in Mendoza as urban development expands into the upland areas of the basin in the foothills. Powerful real estate interests also lobby for these lands (Hurlbert et al. 2015).

Argentina occupies a unique place in **global finance** that impacts national macroeconomic trends. A severe recession in 1998 resulted in the government defaulting on USD 93 billion in external debt, and abandoning the convertibility of its currency on the world market in December 2001; the recession become critical in 2002; foreign investment fled the country in 2002/03; pension funds were nationalized and the agricultural sector became exposed to many price variations, volume swings, and high inflation, all of which continue today (Mauldin 2013; Mussetta 2013). These combine with increasing trade liberalization to exacerbate the vulnerability of agricultural producers.

Informal water practices are more important than formal law in Mendoza. Although the water governance framework governs water rights, interviewees acknowledged that producers relied on social ties with the Inspectors and *Tomeros* in relation to the delivery, timing, duration and quantity of water delivered to their land (M1, M2, M3). The most cited example of this neo-colonial elitism was DGI granting groundwater access and licensing to large agricultural producers. Twenty-one groundwater wells were granted in Tunuyan, Uco Valley by the Governor of DGI to a large producer funded by foreign investors in contravention of groundwater conservation laws (M1). The above drivers are summarized in Table 8.1.

Table 8.1 Drivers impacting agricultural producers' capitals

	Driver – direct		Driver – indirect	
Category	Local	National	Local	National
Demographic	Urban migration and growing population	–	Urbanization	Global population growth
	Shortage of farm labour		Aging producers	Neoliberal market with neo-colonial elitism
Economic	Escalating land prices	Expanding mining	Aging infrastructure	Global price exposure
Political Economy	Increasing cost of inputs, inflation, cost of credit	National monetary policy leads to high interest and inflation rates, lack of capital	–	Increasing trade liberalization in some sectors
				Increasing demand for energy
				Growing inequality
Social	Informal social practices, especially relating to water	–	–	Priority of economy over environment
Natural	Climate change	–	–	–
	Deteriorating ecosystem services			

8.4 Institutions (Organizations) That Build Capacity for Climate Change, d&f

Argentina is a federal republic. Mendoza is a province with its own constitution, laws, and authorities (Mussetta 2013). There is weak federal government presence in Mendoza. Provincial institutions predominate with little or no support from the federal government (Moñtana 2012a). The Climate Change Agency in Mendoza (CCA) and the Agriculture and Climate Contingency Directorate (ACCD) address **climate change issues (ibid)**. The provincial Ministry of Agriculture provides support to agricultural producers (M1).

In 2001 the federal government took a sectoral approach to emergency (SUCCE) but this reduced the capacity of **disaster response** (Pochat et al. 2006). No interviewees could remember this institution being involved in an emergency in Mendoza. Argentinian master **flood** plans do not include Mendoza (*ibid.*). The Argentina National Meteorological Service provides real time information on precipitation and temperature and an online warning system (SMN 2015). The municipalities provide social and emergency services, but are the least economically empowered (M1); as a result municipalities have been creating micro regions (including Mendoza Del Nevado, Lujan de Cuyo y Maipu-Mendoza, and Valle de

Uco) banding together in cooperation to respond to emergency (Altschuler 2012; M5).

The predominant **water** institution in Mendoza is *Departamento General de Irrigacion (*DGI). The Governor of Mendoza appoints and the provincial Senate ratifies DGI's superintendent, executive, tribunal and appeals council members (Mussetta 2013; Diaz and Bertranou 2004). DGI has autonomy to both manage its own resources and create its own rules, separate and apart from the provincial government (M1). The DGI can declare a state of drought.

An intricate water governance system exists that includes CSOs and NGOs. Users of an irrigation canal form self-funded Riverbed Inspectorates responsible for the management of the irrigation network (Diaz and Bertranou 2004). Representatives are elected by votes chosen by rights holders (M6) and water rights holders form DGI's General Users Assembly. Real participation in these user groups happens through social networks in accordance with producer's economic, cultural and social capital (Bustos et al. 2008). These power dynamics influence DGI in relation to urban land use (expansion of the oasis) and land use changes (Mussetta 2013). The DGI influence on water management overshadows initiatives in Basin Councils.

This system also links with an extensive producer network of cooperatives and organizations and a formalized interconnection with government. A 'Family Agriculture Rural Development Provincial Meeting' occurs regularly when the Family Agriculture Provincial Forum, provincial governmental representatives, and public decentralized NGOs meet and discuss concerns (M1). In Mendoza a few water cooperatives provide water supply, however the Obras Sanitarias de Mendoza, a private company, is by far the largest water and sanitation service provider (Diaz and Bertranou 2004).

8.5 Instruments Responding to Climate Change, d&f

Mendoza has a deep, complex system of co-managed water as well as agricultural producer cooperatives. This and the water property interest of 'inherence' are the most important institutions. The main instruments are summarized in Table 8.4.

8.5.1 Regulatory Instruments

Mendoza's water system is highly regulated. However, there are sparse regulations concerning climate change or flood. Water legislation in Mendoza grants the interest in water to the owner of the ground where it is located; as such, an inherence instrument of water allocation is in effect; water can't be sold separately from land (Argentina Constitution 1916, Articles 186–196). DGI oversees the complex regulations of water rights (definitive, eventual, and private), precarious rights (temporal, ground-water, discharge), and priorities (see Table 8.2). Water (and

Table 8.2 Institutional water instruments

Principle	Description
Principle under which water is managed	Public Good (Lee 1990; Baars et al. 1995)
Allocation of water rights	Inherence – Water rights are inherent to the land. Even if the owner doesn't pay the tariff, the right continues with the land
	Licensed interests allocated by DGI on conditions considered appropriate.
Priorities	Human use, irrigation, industry and then fishing and plant ponds
Water market	None
Water pricing	Based on property allocation, not on actual volumetric usage
Water allocation dispute resolution	*Tomero* – Inspector – *Consejo de Aeplaciones* (Appeals Council) of DGI (M2)
Potable water accountability	Drinking water a sewage quality set by EPAS. Service providers bear responsibility
Governance accountability	Although oversight of the DGI is the Provincial Senate, the user associations are governed based on voting and only those with water rights may vote (M2, M6)
Water price	The charges for water are based on an administrative supporting fee, which varies depending on use or source. Profitability is the general criteria (Mussetta 2013)

groundwater) is managed in a supply management structured Q1 fashion (see Pinto 2001).

DGI did not have jurisdiction over groundwater until 1994. However, actual regulations were adopted many years later (Mussetta 2013). Laws 4035 and 4036 relate to groundwater, extraction restraint, suspension, volume measurement and hydro-ecological protected areas. DGI's registry of private water rights is completely out of date (groundwater hasn't been updated since 2002 because of lack of funding) (Mussetta 2013).

Article 162 of the Water Law states that in periods of extraordinary water shortage (**drought**) a system of use by turns (*turno*) is adopted (Diaz and Bertranou 2004). Effectively there is less water provided to irrigators pursuant to their water licences (M2). Inspectors and *tomeros* reduce water supply proportionally (M3). 2014 was the 4th consecutive year of DGI declaring a water emergency because of the severe **drought** reducing water for all uses (Hurlbert et al. 2015).

Argentina does not have specific legislation on **climate change**, an unstructured Q3 problem (Hurlbert et al. 2015). Argentina's needs tend to be met through investment, insurance, and technology transfer (UNFCCC 2015). There were few international instruments present in Mendoza.[1] Similarly, the CCA and ACCD have not taken any direct role in climate **mitigation and adaptation**, but instead an indirect role by encouraging agricultural development and providing advice on avoiding hail and frost (Hurlbert et al. 2015). Several regulatory initiatives that relate to climate

[1] Many capacity building initiatives in Argentina are reported (ibid.), but none in Mendoza. Only two FAO initiatives were identified in Mendoza, discussed below in economic instruments.

change exist: (1) A National Programme for Rational Use of Energy and Energy Efficiency was created in 2005 that encourages the use of bio-ethanol and biodiesel (Pochat et al. 2006; UN 2011). (2) The Glaciers Preservation Law N. 32.016 dictates minimum budgets to protect national glacial water sources (Hurlbert et al. 2015). (3) Argentina's Constitution provides for rights to an adequate standard of living, life, health, food, drinking water, education, housing, work and social inclusion, and a healthy environment (WaterLex and Wash United 2014). (4) Argentina signed the UNCCD (1994) in 1994, prepared a desertification report and the study region of Mendoza is under observation (Pardo n.d.). The national government has instruments responsible for emergency response in relation to **flood**, but no examples of their use were discovered in this research in Mendoza (see Sect. 8.3).

8.5.2 Economic Instruments

National initiatives on **climate change**, exist, but none in Mendoza. The Argentinean Carbon Fund was created to maximize Argentina's participation in the international carbon market through development of CDM projects (Pochat et al. 2006; SAyDS 2008) responding to climate change in a structured Q1 manner.

Although not framed specifically in relation to **adaptation**, several farm programmes alleviate changes in farm income that may have resulted from d&f, or such things as changing commodity prices, structured Q1 problems. These programmes include: the Agricultural Social Programme (PSA) which provides technical and financial support (MA 2015); Provincial Agricultural Services Programme (PROSAP) which funds initiatives that promote the competitiveness of small and medium agricultural producers (PROSAP 2014); credit through the Overhaul and Growth Fund (Mussetta 2013); the Agricultural Solidarity Fund provides crop insurance (DACC 2014); and there is emergency relief (DACC 2014). Private insurance is available, but unaffordable for many (McCord et al. 2013) and there are occasional declarations of tax relief (M1). From the late 1960s, a tariff subsidy on electricity to power the well pump and a tax rebate for building wells and acquiring pumps for ground water have existed. These instruments encourage the accessing of groundwater as an **adaptation** to reduced stream flows (Diaz and Bertranou 2004).

Two international economic instruments were identified in the secondary resources review: A Telefood technical cooperation project promotes awareness and campaigns to increase the capacity of poor farmers to provide food for their families (TFD13Arg001) and a project of institutional investment development in Mendoza (UTF/ARG015/ARG) which reviews PROSAP to improve processes and improve public investment in the agricultural sector (FAO 2015).

Surplus water (of definitive or eventual rights) can be allocated annually through a process of re-appropriation and reallocation (Mussetta 2013). There are no legislated criteria for allocation or reallocation based on benefits stemming from water use (i.e. value of production) and the incentive to do so is only a tax credit (Diaz and Bertranou 2004). No interviewee recollected this instrument being accessed.

8.5.3 Suasive Instruments

A few suasive instruments exist in relation to climate change and water. A Climate Change National Strategy (CCNS) sets out a strategy for **mitigation and adaptation** including a number of action guidelines for programmes, projects and indicators (Nazareno 2013). However, little is implemented in practice (Hurlbert et al. 2015). The Leading Principles of Water Policy (*Principios Rectores de Politica Hidrica*) sets out common rules and regulations of all Argentinian provinces in respect of water and although there has been discussion to codify this into federal Argentinian law, this has not yet occurred (Pochat et al. 2006). In Mendoza, the provincial CCA defines its priority as promoting energy and water efficiency through audits, diagnosis and awareness campaigns, and small Q1 programmes.

8.5.4 Managerial Instruments

The Water Users Assembly, *juntas* and *tomeros* implement a complex system of supply and co-management of water for water rights holders employing managerial instruments. This system of co-management is located in Q1 at the top left side where water issues are structured and the system has little flexibility to respond to uncertainty in norms and values. The co-management process involves setting fees and electing representatives. Significant in its absence are transitionary insolvency provisions such that outstanding loans and debts are never discharged (Martindale-Hubbell 2001).

Other managerial instruments include: (1) a Master Plan for Mendoza River Basin developed in 2010–2012 in respect of five river basins after only one focus group meeting (M6) therefore residing as a policy problem in Q4. (2) a Q4 DGI water plan 'H2020' covers water infrastructure upgrades, legal and institutional changes; no mention is made of climate change and adaptation (DGI 2015). (3) an integrated land use plan was embarked upon in 2014 (Planning Law 8051) through public consultations. However, the plan would be subject to the water laws of Mendoza, thus reducing the potential impact of this exercise (M4) to address unstructured problems.

8.6 Adaptive Governance and Problem Structuring

Figure 8.2 depicts how the instruments responding to d&f in Mendoza are characterized mostly as Q1 structured problems. However, significant Q4 and a few Q3 examples were ascertained. Mendoza's policy is predominantly structured around technocratic Q1 problems of stabilizing income, emergency relief, or tax reductions (Mussetta 2013). In response to drought a technocratic Q1 process of '*turno*' is implemented.

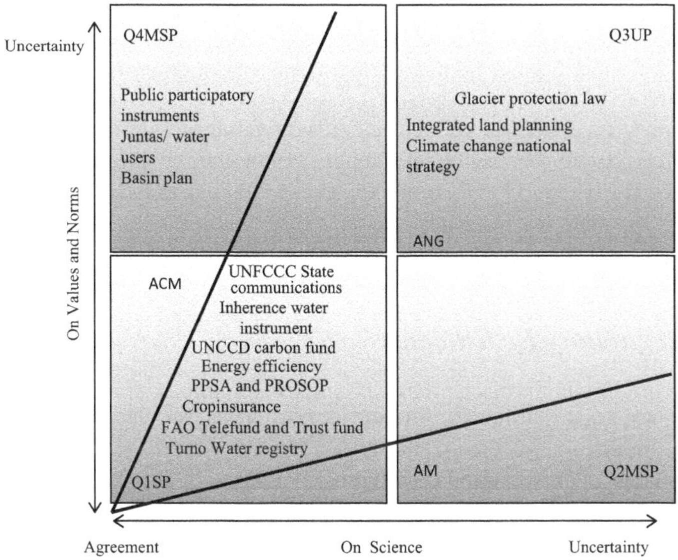

Fig. 8.2 Mendoza's instruments and problem structuring
ACM Adaptive co-management, *Q4ACM* Quadrant 4 moderately structured problem, *ANG* Anticipatory governance, *Q3UP* Quadrant 3 unstructured problem, *Q2MSP* Quadrant 2 moderately structured problem, *AM* Adaptive management, *Q1SP* Quadrant 1 structured problem

Emergency planning does not include climate change, d&f and is not part of the emergency psyche of local Mendocinas; although interviewees were all well versed in issues of uncertainty and climate change, the discursive analysis of the policies did not establish a theme of uncertainty and climate change. One interviewee stated that urban residents are not aware of the scarcity of water having lived in an oasis that has existed for centuries and are ignorant of the Declaration of Emergency over the past 4 years (M4). Another stated, "The consequences of climate change are here, but the provincial authority, DGI is not taking them seriously" (M7).

Although a national Climate Change Strategy exists and there is a Mendoza office to facilitate adaptation (CCA)(Q3 measures), implementation is yet to occur (Hurlbert et al. 2015). Mendoza's Glacier Protection Law is a good example of a Q3 instrument (M1). The PRODESAL and INDAP programs are targeted directly at climate change adaptation, a Q3 problem.

Environmental governance approaches of adaptive co-management (Q1 and Q4) are part of the institutional fabric of Mendoza, but only for those with water interests (which are attached to certain property titles). The rigid structure of inherent water interests impedes the ability to respond to uncertainty as no changes can be made in the proportion of water delivered to an irrigator (M1). Local water groups dealt with moderately structured issues (such as what fees and charges should be paid) and water practices deviating from the rules of inherence. Although public consultations over land use planning, water, and environment are common, no other forms of environmental governance approaches were found facilitated by government. No Q2 adaptive management instruments were uncovered.

8.7 Impacts of Instruments on Actors Measured by Mandate Effectiveness

The regulatory inherence instrument significantly impacts agricultural producers, supported by a strong practice of co-management of water within irrigation associations and the DGI; economic instruments are also increasingly important. Table 8.4 summarizes the main instruments.

8.7.1 Regulatory Instruments

Argentina has no credible **climate mitigation** measures and its emissions are increasing slightly, as expected from the developing countries (Di Paola and Rivera 2012). The Glacier Preservation Law indirectly addresses climate change and preservation of glaciers by protecting water for viticulture and keeping mining from accessing the glacier (M1). No **adaptation** plans are in place and there are no measures to reduce vulnerability in the long term (Hurlbert et al. 2015; M6); vulnerable people are still at risk (Vos et al. 2010).

The Argentina Supreme Court has ordered the government to recognize constitutional rights to **water** and a healthy environment (WaterLex and Wash United 2014). Regulatory instruments have been successful in this area. Interviewees supported federal government initiatives surrounding access to water and sanitation, and WHO statistics confirm improvements with 27% and 29% of the population gaining access to clean drinking water and sanitation facilities respectively since 1990 (2015).

Regulatory instruments are efficient in allocating water and supporting irrigated agriculture. However, some issues exist which prevent adaptation. First, the licensing of groundwater by DGI to powerful agricultural producers is depleting aquifers (M1). Second, the inherence instrument of water allocation does not allow flexibility to adapt to changing livelihoods; the process of *turno* is entirely supply managed with no flexibility to respond to drought. The *turno* instrument does not allow crop location to be changed, water shared, or water permanently transferred to another use. If water fees are not paid, the owner does not get to vote in water governance; however, the water interest stays with the land, and the right is not lost (M2). This tariff acts as a shield for small powerless agricultural landowners as it prevents them from losing land and water rights. Groundwater regulatory instruments are less efficient. Groundwater pumping is managed by DGI in a centralized, non-transparent, and discretionary manner (Montaña and Boninsegna 2016).

Water governance is characterised by sectoral fragmentation; it lacks inter-institutional coordination, communication and information exchange between actors (Vega 2011). Inter-sectoral conflicts result (especially between competitive uses such as irrigation and hydroelectric energy generation) over flow quotas. There is also poor management of floodwater and contamination of interprovincial water courses (Pochat et al. 2006).

8.7.2 Economic Instruments

Economic income stabilization instruments (insurance, credit etc.) have effectively supported many irrigated producers (M2). However, often, these are not accessible by small producers (M2; M6) and when these producers get some kind of aid, it is never enough to recover from damages (M6) as they are only a small payment to replace lost inputs or a tax waiver (Mwinjaka 2008). The two programs aimed at small farmers, the ASP and PROSAP, can only be accessed if water charges are not in arrears; this excludes many (M7). Even with the tariff rebate on electricity, the current high cost of electricity necessary for pumping groundwater, has (along with the regulatory instruments surrounding groundwater) made such instruments inaccessible for most medium and small producers (M2).

Economic instruments don't adequately address the driver of Argentina's monetary situation. Small horticulturalists (producing fruit and vegetables, or grapes for sale to the four large purchasing companies) are impacted more fluctuating market prices for their produce and exposure to financial markets. This driver's significance overshadows the availability of economic instruments and even the process of *turno* due to drought (M2).

8.7.3 Suasive Instruments

The activities of the CCA and ACCD have had little impact on agricultural producers (M1). **Climate change** plans developed either by Argentina, or by Mendoza, have little bearing on the management and use of water; climate and water are two distinct institutional silos (M4). Initiatives such as the Argentinian Leading Principles of Water Policy, H2020, and even the proposed **water** registry that might advance climate adaptation in relation to water have little relevance to the real institutional water governance system (M2). The lack of a finalized registry of water makes management of water difficult and not transparent (M6).

8.7.4 Managerial Instruments

The sectoral emergency planning of SUCCE has not responded well to **flood**. One flood outside of Mendoza washed out a major highway linking Mendoza to the capital of Argentina for several days. The people involved in this incident had no communication, and did not receive a response for seven days (even given the organizations with emergency mandates) (M2).

Participation in **water** governance via new institutions such as water basins are not believed to significantly impact water governance or climate adaptation in Mendoza. The DGI, the *Inspectorates* and *tomeros* are powerful (M1, M6) which limits change. Even within the General User Assembly, voting procedures aren't conducive to fair representation or **adaptation to d&f**. Those without water rights are unable to vote. Downstream individuals and those involved in goat husbandry have no input into the water governance of Mendoza (M6). Argentina has no integrated water management, partly due to institutional rigidity, fragmentation and narrow mandates (M6).

8.8 Assessment of Learning

Mendoza has the greatest depth and breadth of learning of all the case studies. In chronological order, triple loop learning occurred historically with the creation of the oasis of Mendoza from the dessert in the sixteenth century. Now, however, the Mendocina society is organized around the common cause of 'beating the desert' that competes against a sustainable land model (Montaña et al. 2005: 5; M2). Successive expansion has occurred with the last phase in the Uco valley by large local and foreign enterprises requiring intensive capital, utilizing foreign genetic plants, modern technologies of cultivation and irrigation (*ibid*).

The Potrerillos dam of 2002 is an example of zero loop learning. This dam modified the natural flow of Mendoza River to benefit irrigators (as it improved their adaptive capacity). Although there was a political discourse that irrigators and government would increase irrigation efficiency to provide more water to dryland farmers, the dam prevents any water reaching the dryland (M5). This weakened the position of dryland farmers who previously might have received some water during peak snow melt for goats, albeit never enough to irrigate and produce crops (M5).

The successful lobbying by Mendoza producer groups and environmental groups (the Popular Water Assembly) and establishment of the Glacier Protection Act shows triple loop learning for a Q4 problem as it effectively prevents mining development for economic growth at the headwaters of the Mendoza river. This development would have detrimentally impacted already constrained water resources used by downstream irrigated agricultural producers and the Mendoza oasis (M1).

Examples of single loop learning include medium sized irrigators improving the efficiency of their operations. By accessing credit instruments these irrigators are adopting practices that reduce the amount of water lost during its delivery from the dam to the farm (by lining and covering canals) (M2) (Table 8.3).

Table 8.3 D&F: Assessment of learning in Mendoza

Level/learning evidenced	**Change in practice – single loop learning** (one way information flow – Q1)	**Change in assumptions or models – double loop learning** (two way information flow – Q2 Q4)	**Change in values, norms, world views or power dynamics – triple loop learning** (iterative information flow – Q3)
Individual	Adoption of efficiency measures	–	Glacier Preservation Law
Regional		–	
National	–	–	–

8.9 Instrument Impacts on Livelihood Capitals

Social capital is very strong in Mendoza, although different dimensions exist for different sizes of producers. The irrigation and producer organizations as well as iterative public consultations (on water governance, glaciers, integrated land management, etc.) contribute to this. Size of agricultural producer and scale of production impact the access to economic, technological and natural capital (see Table 8.4).

8.9.1 Human Capital

A large discrepancy exists regarding access to education and health services between rural and urban people, and it is worsening. Although cooperative groups are trying to maintain young people on the farm (M2), increasing urbanization and aging producers impacts Mendoza. Urban dwellers have expanded by 40% from 1986 to 1999, while rural residents have remained constant (Mussetta 2013). There are less *Campesinos* on the land than in the 1990s and they tend to be poorer; in times of drought the animals die, and the people migrate (M5).

8.9.2 Social Capital

In Argentina (as in Chile) there has been a growing stratification between agricultural producers in the last few decades with implications for social, economic, and technological capitals. Twenty years ago producers were all relatively small or medium sized. Now there is a significant group of large producers (M3) who understand and differentiate between various state levels and institutions, knowing which state bureaucracy (offices and functions) to access; are better able to access economic benefits (tax breaks, direct subsidies, state grants) and technical agencies assist them with making informed decisions on issues such as water supply, efficiency, and financing. Small producers cannot differentiate different state

Table 8.4 Assessment of main instruments

Type	Instrument	Eff	+ Impact on capitals	– Impact on capitals
Reg	Inherence water instrument	–		– Economic
	Water Tariff based on land	–		– Economic
	Declaration of Drought – *Turno*	–		– Economic
	Glacier Preservation Law	+	+ Natural/Social	
But miss	Instruments to promote efficient water use and irrigation			
	Climate change mitigation and adaptation tools			
Econ	Groundwater rebate	–		– Natural
	Loan instruments	+	+ Economic	
	Agricultural solidarity fund	+	+ Economic	
	PSA	+	+ Economic	
	PROSAP	+	+ Economic	
	Emergency relief	+	+ Economic	
But miss	Flood insurance			
	Debtor protection and forgiveness in bankruptcy			
	Water transfers			
	Irrigation expansion incentives			
Suas	Leading Principles of Water Policy	–	–	–
	Climate Change National Strategy	–		
But miss	Environmental stewardship incentives			
Mgt	Master Plan for Mendoza River Basin	–		
	Water user assembly, juntas, tomeros co-management	+	+ Social	–
	Water Registry	–		
But miss	Drought and Emergency Flood planning			
	Inclusive, participatory development instruments			

Inst Instrument, *Reg* Regulatory, *Miss* Missing, *Econ* Economic, *Suas* Suasive, *Mgt* Managerial
++ Effective, + Moderately effective, – Ineffective

institutions, offices and functions except for the municipal level from which they may receive immediate assistance (especially in extreme climate event situations) (Hurlbert et al. 2015; M3).

Large producers use the same conflict resolution mechanisms as small and medium producers through their *tomero*, the inspector, but in addition can access the Honourary Tribunal of DGI, and the court system (in the event their interests are not satisfied by the traditional methods (M3)). One example cited was the granting by DGI of an underground water licence for 21 wells in the Valley of Uco to a large foreign producer and winery in contravention of a groundwater conservation law. The producer was able to continue to utilize groundwater by claiming the rights granted contrary to the legislation in court and continuing to access groundwater as the court case proceeds (M4).

Small grape producers with traditional vineyards and horticulturalists exist generally downstream. The horticulturalists resort to social and family networks to organize their production and successfully develop their agricultural activities. In times of water shortage, small and medium producers, without access to groundwater, rely on relations with (and payment to) the *tomero* and inspector to receive water. It is possible for a *tomero* to provide water to a producer who is in dire circumstances and at risk of losing vines or fruit trees (M1, M2).

Downstream dryland goat herders and horticulturalists, *Guarpes*, "never had water, and now have even less" as a result of the building of the Portrerillos dam (A5). These poorer people are described as having strong social capital within their local communities (the providers of emergency assistance), but having no links with institutions outside their local communities (M5). As an example, during the participatory integrated land planning process of 2014 the *Guarpes* were not at the negotiating table (M4). Development such as the building of the Portrerillos Dam occurs often at their expense (M5). These small producers don't qualify to access financial tools, which requires them to have all taxes and fees being paid up to date (M5).

8.9.3 Economic Capital

Large producers have high economic capital and adaptive capacity. They capture the profit involved in making wine and can weather drought by access to groundwater. Unlike large producers, medium and small producers are unable to realize the profit between the price paid for the grapes and the price paid for a bottle of wine and cannot access groundwater (Diaz and Bertranou 2004).[2] Large producers have a further four advantages in relation to economic capital. First, the 10–15 largest producers have market dominance and an export trademark for the wine they produce and wholesale from smaller producers frequently with foreign investment. The second category of large wine producers (55–60) and all medium producers sell their grapes to these large producers that produce common wines that include Trapiche, Bajou, and Folita (M1).

Second, the DGI has utilized its instrument of forfeiture for non-payment of water fees in two of the most expensive areas (Uco Valley and to the right margin of the Luhan River) on behalf of large producers (M3). The DGI has ejected these small producers making way for large foreign funded producers; in other oasis areas forfeiture for non-payment of water fees is unheard of (M1).

Third, the power of large producers is enhanced by the criteria of water priorities in Article 116 of the Water Law which allows, within each class of water rights holder, business of 'greater importance and utility' or larger more profitable businesses to be given priority (Diaz and Bertranou 2004: 43). Fourth, concessions for

[2] Most irrigated producers are small producers, with farms less than 10 hectares; many producers have medium sized farms from 10 to 40 hectares; few (approximately 70 with 10–15 being significantly larger) have farms ranging from 700 to 1000 hectares (M2).

ground water are granted only by DGI and have only been granted to the larger producers (Mussetta 2013; Diaz and Bertranou 2004).

A few instruments assist medium-size producers. Economic instruments help stabilize income and new credit facilities are increasing access to technology (M2). The regulatory instrument of inherence preserves the economic capital of small and medium producers; if they refuse to sell their land, their rights are preserved and the concentration or monopolisation of water rights by one owner restricted (Diaz and Bertranou 2004). Other than in the two areas where forfeiture has been practised to the detriment of small producers (M3), producers hold on to their water rights (M2).

For small and medium producers often employment off farm (working in wineries) is required to support their livelihood, especially in times of drought. These small and medium producers may also rent out their land for a few years in order to pay off their debts before returning to their production activities (M2).

8.9.4 Technological Capital

Irrigators have benefited from investment in infrastructure to deal with water shortages including the building of the Portrerellos dam (Montaña and Boninsegna 2016); the goats of dryland Campesinas have been negatively impacted.

Access to information about new technologies, climate change, and social programmes are unequally distributed. Large producers have the most channels; they can obtain information from vendors of goods, INTA, large institutions such as DGI, and research institutes, cooperatives (they may belong to), and they can hire consultants and specialists. Medium and small producers have some of these avenues, but have no access to large institutions like DGI; the smallest producers will not even have access to cooperative associations, but rely on INTA (M2).

Large producers use innovative strategies in response to climate including automatic irrigation systems (drip irrigation and others) and large capitalized producers sustain their agribusiness with access to groundwater (Montaña and Boninsegna 2016). This technology enables these producers to diversify their locations and relocate properties in the foothills upstream. Medium and large producers increase their efficiency through improved technology by accessing credit. Often this entails the lining and covering of water channels, a form of adaptation (M2). These types of adaptations are unavailable to small producers because of lack of access to funds (Hannis 1977).

8.9.5 Natural Capital

Natural capital in Mendoza is deteriorating. There is pollution and salinization of land. The national environmental law and the Environmental Federal Council (COFEMA) are ineffectual as resource limitations restrain adequate activities and enforcement (Salerno 2009; Cetrangolo et al. 2004; Hurlbert et al. 2015). Industries pollute water by discharging effluents into irrigation canals and transferring these externalities to users downstream damaging both health and crops. In some areas, the poor control of water for cleaning in the oil industry has directly resulted in polluted aquifers (Diaz and Bertranou 2004).

Floods affect properties in the Mendoza River basin. Inadequate drainage decreases yields due to soil depletion, build-up of clay silt at the foot of irrigated rows, and increased salinization (Hansis 1977). The Potrerillos Dam aggravates salt accumulation in the Northern Oasis (Diaz and Bertranou 2004). The continuous expansion of urban sprawl and the relocation of large capitalized producers to the foothills upstream are threatening aquifer conservation and the agro-ecological conditions of downstream lands (Montaña 2012a, b). (This process was worsened by the electrical tariff subsidy and tax rebate for groundwater development (Diaz and Bertranou 2004)).

The Climate Change National Strategy does not address this reduction in natural capital, but instead focuses on how to adapt the current model of production to new climate conditions (M1). The focus is not on limiting the expansion of the agricultural frontier, thereby limiting deforestation and soil degradation (Hurlbert et al. 2015). It also does not seek to promote agri-ecological practices (*ibid.*) (Table 8.4).

8.10 Re-designing Instruments with the ACW

This chapter explored how Mendoza's d&f policy framework impacts the capitals of agricultural producers, given drivers. I use these findings to assess the adaptive capacity of Mendoza in the ACW (see Fig. 8.3) and then suggest design improvements to enhance the adaptive capacity of the governance institutions of climate change, d&f, together with conclusions surrounding adaptive governance.

There is a deficit of **variety** as problems of climate change, d&f all lack instruments (see Fig. 8.2) reducing problem frames and solutions. Although currently a public consultation on land use planning is occurring, tackling moderately structured problems raised by environmentalists (M4), this is a short-term process. The Argentinian state is largely absent, reducing the number of actors and instruments.

The irrigated oasis is the overarching reality (M3) historically resulting in triple loop learning; other drivers such as climate change, d&f aren't cognized by the public. This has limited the **learning capacity** of producers and stymies single and double loop learning (such as adoption of efficient drip irrigation technology). The predominant political model is akin to a state welfare system (as demonstrated by the economic instruments for small producers and failure to end water rights for non payment of fees); the ravages of the neoliberal state model have yet to be experienced (Montaña and Boninsegna 2016). Difficulty is experienced with dwindling state **resources** (*ibid.*). Nevertheless, strong leadership of government and civil society (including irrigated producers) and **autonomous change** in Mendoza contributed to the triple loop learning and collaborative **leadership** reflected in the

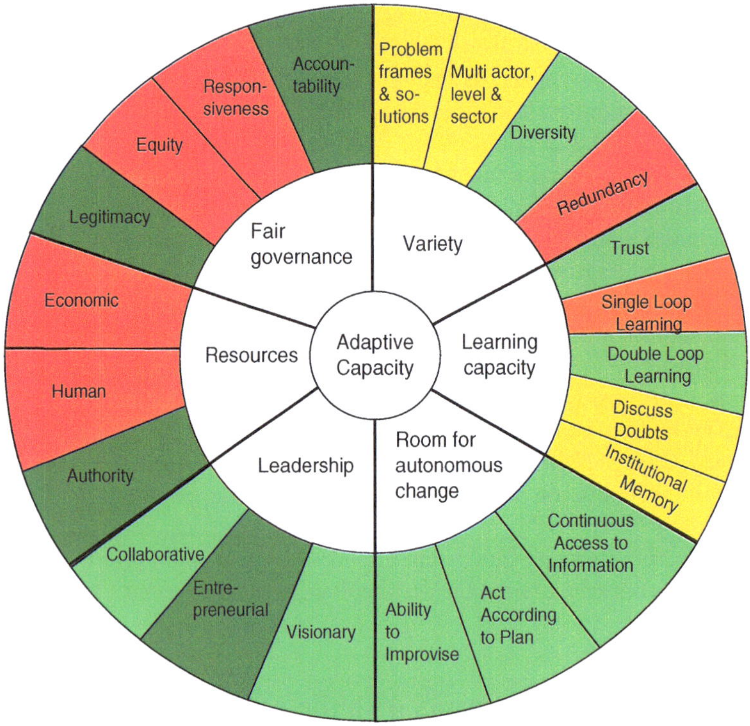

Green	Light Green	Yellow	Light Orange	Red
Very High	High	Medium	Low	Very Low

Fig. 8.3 Mendoza's ACW

passing of the Glacier Protection Law, which prevents expansion of mining and its associated GHG emissions, and preserves the livelihoods of the irrigated producers by retaining Mendoza water for agricultural use.

The irrigated oasis and the inherence water instrument have created a built environment that protects small and medium irrigators, but also limits their room for improvising, **autonomous change, and entrepreneurial leadership**. The inherence principle prevents adaptation as water delivered to producers is proportionally reduced without any attention to water needs. The fragmented nature of dealing with water has resulted in a deficit in relation to technologies such as capturing rainfall, making better use of groundwater, and fostering the combined use of surface/groundwater.

Resources are strained partly due to the drivers of government austerity, neoliberalism and the neo-colonial state. The government's monetary policy has a large impact on Mendoza and its producers, but in areas such as disaster response and climate change policy, the national government is largely absent. Large producers have access to resources including economic, technological and natural capital, but not small ones.

In respect of **fair governance**, an elaborate institutional system of water governance through the General Users Assembly and Riverbed Inspectorates advances **responsiveness, accountability, legitimacy, and collaborative leadership**. However, these governance processes don't include people without water rights giving rise to issues of **equity**. Large capitalized producers who are situated upstream of the basin (thus able to preferentially access water) are more adaptive and resilient in the face of drought, whereas smaller producers at the tail end of the system are receiving less water and have less access to capital to adapt to drought (Montaña and Boninsegna 2016). The exodus of these subsistence producers is tempered by the availability of economic instruments and emergency relief, and the lack of insolvency and bankruptcy provisions available to them.

Redesign options for Mendoza include addressing the drivers of the national monetary policy and increasing trade liberalization, demographic changes including urbanization, and growing inequality. Granting access to medium and small producers to the same marketing opportunities of large producers by setting aside neocolonial practices, allowing equitable access to groundwater, technology, and economic capital would create a fairer playing field. Missing instruments such as promoting efficient water use and irrigation, responding to flood, limited water transfers, environmental stewardship incentives need to be developed. Strengthening the ineffective instruments, especially modifying the water inherence principle so that it allows for adaptation, would increase the adaptive capacity of rural agricultural producers.

Changes need to occur while preserving: (1) the customary practices of water occurring at the riverbed inspectorate and tomero level allowing flexibility of water delivery on a local scale; and (2) the leadership and social capital that has contributed to the single and triple loop learning in Mendoza. For irrigated producers, the water inherence and economic instruments have created a hydraulic society that should also be recognized and preserved (see also Boelens and Vos 2014 for a detailed account of Andean water practices in another area). The built infrastructure of irrigation has historically protected the livlihoods of agricultural producers as they have continued to retain their land and water rights even during years of non-payent of water right fees (but this is starting to change in areas highly sought by large producers). Although Mendoza has a history of utilizing iterative participatory instruments that needs to be preserved and replicated, this process needs to include meaningfully those without water rights, the *Guarpes* (Table 8.5).

The Mendoza case study confirms the importance of people, civil society groups and political leaders in leadership roles to advance triple loop learning and envision long-term transitions to sustainability via initiatives such as the Glacier Protection Law. No doubt the strong Mendocinian social capital as well as the culture and built infrastructure of the 'oasis' were important factors in this demonstration of adaptive governance.

Table 8.5 Redesign options for Mendoza

Analysis Unit	Description
Ineffective instruments	Flood measures
	Climate change mitigation and adaptation plans, (water, frost hail information provision), carbon fund
	Glacier preservation law
	Water tariff, well licensing, groundwater regulating, inherence water instrument, leading principles of water policy
	Master plan of Mendoza river
Missing instruments	Water transfers
	Environmental stewardship
	Drought and emergency planning
	Flood insurance
	Efficient water and irrigation practise
	Insolvency
Unaddressed drivers	Government austerity (national monetary policy)
	Increasing trade liberalization
	Demographic urbanization, population grown
	Growing inequality
	Deteriorating ecosystems
Weak adaptive capacity dimension	Fair governance (equity)
	Variety
	Resources (economic and human)
Constrained capitals	Human (discrepancies of poor)
	Economic (small producers impacting access to technological and natural)

References

Altschuler B (2012) Fronteras sociales y asimetrías en la vitivinicultura mendocina actual. Cuadernos de desarrollo rural 9(68):151–175

Baars E, Bastiaansen APM, Menenti M (1995) A user-oriented and quantifiable approach to irrigation design. Water Resour Manag 9:95–113

Boelens R, Vos J (2014) Legal pluralism, hydraulic property creation and sustainability: the materialized nature of water rights in user-managed systems. Curr Opin Environ Sustain 11:55–62

Boninsegna J, Villalba R (2006) Los condicionantes geográficos y climáticos. Documento marco sobre la oferta hídrica en los oasis de riego de Mendoza y San Juan. Primer informe a la Secretaria de Ambiente y Desarrollo Sustentable de la Nación. p 19

Bustos RM, Saldi R, de Rosas L (2008) Quien tiene la ultima palabra? Discurso Institucional vs. Redes sociales en la descentralización de la gestión del sistema de riego en la cuenca del Rio Mendoza. En: Tmas y debates 15/dossier/agosto 2008, pp 83–96

Cetrángolo O, Chidiak M, Curcio J, Gutman V (2004) Política y gestión ambiental en Argentina: gasto y financiamiento. CEPAL, Santiago de Chile

DACC (Direccion de Agricultura y Contingencias Climaticas) (2014) Website. Available at: http://www.contingencias.mendoza.gov.ar. Accessed 30 Nov 2014

DGI (2015) Plan H2020. Available at: www.aqua.gov.ar/2020/. Accessed 21 Jan 2015

price as a dominant risk to their livelihood. This driver has local dimensions including access to these markets. In Canada, institutional atrophy (the loss of marketing agencies such as the Canadian Wheat Board) means producers sell directly into the market via private companies. In Argentina and Chile the neo-liberal and colonial culture results in very large producers selling internationally with only a few selling produce nationally (Hadarits et al. 2016; Montaña et al. forthcoming). The other producers sell produce locally (Montaña and Boninsegna 2016) at much reduced prices.

The second international economic driver is financing or obtaining credit. The size of the agricultural producer's operation and the attainment of credit internationally differentiated very large producers from medium and small agricultural producers. In Mendoza, most producers have difficulty in obtaining credit and paying the related prohibitive interest rate and have difficulty expanding operations, except those with international financing. Argentina monetary policy makes wine and horticultural producers vulnerable to fluctuating exchange rates making exports risky. The monetary policy and state austerity leads to limited and insufficient state aid (Montaña and Boninsegna 2016).

The third economic driver is related to the shortage of farm labour. Many interviewees identified the cause of this shortage as the higher paying urban jobs and jobs within the oil economy. Agricultural producer revenues, and return on agricultural investment cannot compete with these higher incomes. The fourth economic driver present in all countries is the prioritization of the economy over the natural environment. This deflects focus from climate change and the fifth economic driver: growing inequity.

Focus on the economy, and away from the environment and equity inhibits adaptive governance and triple loop learning. Perhaps it explains why the triple loop learnings of the international level (see Sect. 4.6) are absent locally. Considerations of mitigation and adaptation in disaster recovery and development as well as the involvement of more stakeholders and the public in planning can be expensive and time consuming. In Canada, this is exaserbated where climate change is mostly denied, as in Alberta (see Sect. 6.3), or treated with scepticism, as in Saskatchewan (see Sect. 5.3).

Focus on the economy also has meant leaner government and reduced taxes to encourage business growth in the case study areas and associated government austerity. Austerity reduces government social supports, leaving this to the private sector and instruments such as insurance, which are unaffordable to many. Municipalities, with smaller tax bases (growing smaller with urbanization) are left to use limited tax revenue to provide emergency aid, welfare, and relief. Government austerity also means less government programs, programs that in the case studies advanced resilience to d&f.

Table 9.1 Drivers impacting agricultural producers' vulnerability to d&f

Category		Driver – direct		Driver – indirect	
		Local	National	Local	National and global
Demographic	All	Lack of farm labour	–	Urbanization/population growth; Aging producers	Global population increase
	Can	Climate change denial Increasing farm size			–
	LA				Neoliberal market culture Colonial elitism Low transparency, poor enforcement of water law
Economic	All	Escalating land prices Oil and mining development	–	Aging infrastructure (roads, dams)	World Recession 2008 affecting economic development and austerity
	Can	–	–	Reduced size and budget of government	–
	LA	–	–	Strong viticulture sector (Ch8Ar)	Currency fluctuations (Ch8Ar)
Political Economy	All	Increasing cost of inputs	Reduction of government programs	–	Growing inequality Demand for energy
	Can	–			
	LA	–	Monetary policy leads to high interest/inflation and lack of capital (Ch8Mz)		Social supports left to private sector (Ch7Co) Increasing trade liberalization

Social	All	--	–	–	Priority of economy over environment
	Can	–			
	LA	–			
Natural	All	Climate change	–	–	–
		Deteriorating ecosystems and concomitant loss of services (FPTGC 2010)			
	Can	–			
	LA	–			

All All Countries, *Can* Canada, *LA* Latin American countries, *Al* Alberta, *Sk* Saskatchewan, *Co* Coquimbo, Chile, *Mz* Mendoza, Argentina

9.3 Main Formal Institutions Responding to Climate Change, d&f

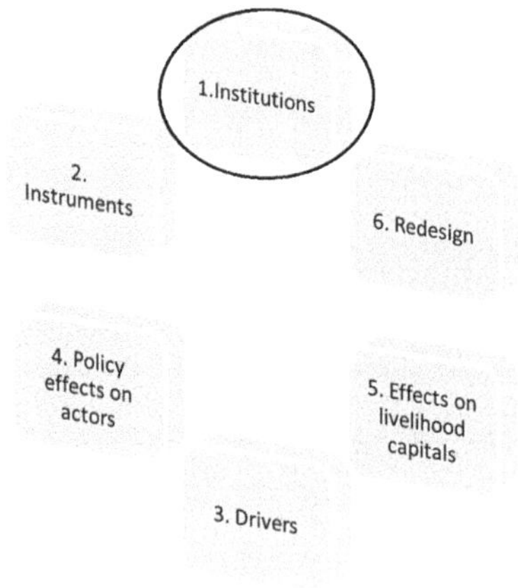

Table 9.2 identifies the main formal institutions or organizations responding to extreme climate events of d&f at the global to regional level of each study area.

This research illustrates the importance of interconnection amongst these institutions, both horizontally and vertically at all levels for governance in order to advance adaptive governance and agricultural producer livelihoods. Although a formal network analysis of these formal institutions was not part of this research, by identifying their instruments present in the case study areas and the linkage with agricultural producers inferences of interconnection can be made. Significant disconnections that impacted agricultural producers exist in relation to international instruments (not being present at the case study level) and in Chile between the strong national government and absence of effective instruments at the case study level. The lack of interconnection is also linked with power differentials between levels of government (Clarvis and Allan 2013; Bauer 1997; Carruthers 2001; Galaz 2003) (see Sects. 7.4 and 8.4).

In Argentina, the strong provincial government lacks funds and funding support from the central government. In Chile the strong central government, together with the lack of a provincial or regional government provide little guidance or financial support to the local municipal governments. Scarce Chilean governmental resources have been allocated to earthquake disaster recovery, leaving the disaster of d&f to local response. Flood has not yet been a significant concern in Mendoza, although heavy rainfall is predicted (IPCC 2014). National and regional emergency response organizations are missing, or lack resources.

Table 9.2 Organizations in study areas responding to climate change, d&f

Level				
International	UN, WTO, World Bank, UNFCCC, IPCC, FAO, Future Earth, UN Water, GWP, IWA IWMA, IWRA, Aqua Fed, AIDA, UNDP, UNISDR, UNEP, Red Crescent and Red Cross, Salvation Army TIEMS, Global Risk-Davos, Reinsurers, Insurance Companies.			
	Study Area			
	SK	**Alta**	**Chile**	**Argentina**
National	Ministries of Agriculture, Environment and Public Safety, Natural Resources Canada, Health Fisheries and Oceans		General Directorate of Water (DGA)	Ministries of Env, economics, production public works
	Environmental, water and safety civil society groups, Ducks Unlimited, Forum of leadership on Water, Canadian Water and Wastewater Association, SSRB Advisory Committee etc.		Hydraulic Works Authority (DOH)	Nat. Institute of Agricultural Technology
			Nat. Irrigation Commission (CNR)	Central Sub Unit of Coordination for Emergency
			National Emergency Office (ONEMI)	
Inter-regional	Government ministerial committees		Climate Change Agri Council	Transboundary river committee
	Prairie Provinces Water Board		Regional Emergency Response Committee	Environmental Federal Council
	Red Cross, Salvation Army			
	Watershed advisory committees			
Regional (Provincial)	Water Agency	Water Ministry	Regional representative of DGA	DGI
	D&f Committee	Drought Management Committee		Agri-industry and technology ministry
	Public Safety Ministry	Emergency Management Agency		General water users assembly
	Environmental/Aid Groups	Environmental/Aid Groups		

(continued)

Table 9.2 (continued)

Local-municipal				
	Municipalities	Municipalities	Municipalities	Municipalities
	Irrigation Districts	Irrigation Districts	*Juntas de Vigilencia*	Riverbed Inspectorates Associations
	Emergency response groups	Emergency response groups	Water communities, user associations (potable water, drainage, agricultural, groundwater)	Inspectors
	Local watershed groups	Local watershed groups	Agribusiness, mining companies, hydroelectric companies	Basin Councils
	Agri-business	Agribusiness		Popular Water Assembly
	Mining	Oil and Gas		Irrigator cooperatives
				Multinational agricultural producers

9.4 Main Instruments Responding to Climate Change, d&f

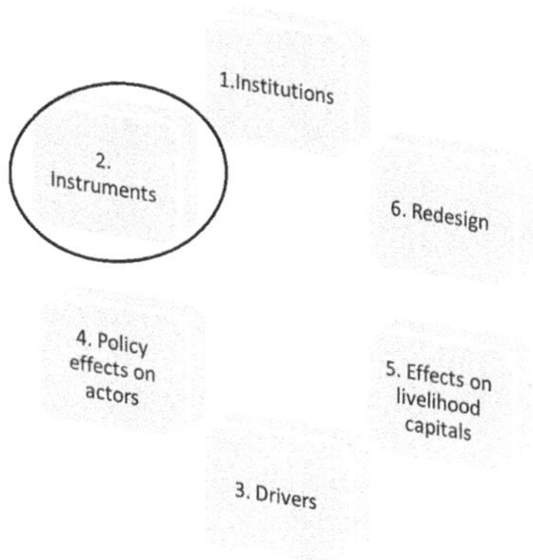

Table 9.3 lists the main instruments organized by the policy problem of water and climate change, and the embedded policy problems of d&f. An indication of the ranking of the effectiveness of the instruments used (see Sect. 3.3.5) appears in a column to the right (discussed in the analysis of impact on actors, see Sect. 9.6). Table 9.3 also identifies missing instruments.

Two categories of missing instruments are notable. First, instruments (especially international ones) that tackle the unstructured Q3 problem of climate change are missing. These include effective GHG emission reduction regulations, and economic instruments (cap and trade, carbon markets or taxes in GHG emissions). Second, there were also no current adaptation policies or instruments within the case study regions.

No government had direct programme spending on research on climate change mitigation and adaptation. Instruments supporting the environment such as payment for ecosystem services, conservation tenders, conservation auctions, or energy and/or water demand management were also missing. There are no instruments dealing with the systemic nature of d&f such as fines for illegal drainage, payments for retention and creation of wetlands for water storage, or demand management of water. There are only a few instruments providing for occasional public participation in relation to climate change, d&f. This lack of participatory instruments is one of the most serious defecits and will be discussed further in Chap. 10.

The DCs have fewer instruments with the least variety and significant gaps in suasive and management instruments. The DCs had few if any instruments responding to flood (although it is anticipated to worsen due to the driver of climate change), lacked significant effective regulatory water instruments, instruments requiring emergency planning, and source water protection planning. Unlike the DCs, Canada had no programs aimed specifically at small or poor producers, no formal water rationing system (although voluntary practices existed), and no glacier protection laws or policies.

Table 9.3 Instruments responding to climate change, d&f

Inst		**Water/Climate change**	**Eff**	**Drought/Lack of Moisture**	**Eff**	**Flood**	**Eff**
Reg	**All**	Environmental liability	–	–		Drainage permits	–
		Water pollution control/Water quality standards	+			Drainage liability	–
		Waste water treatment controls	+			Right to humanitarian aid	–
		Land use restrictions	–			Building standards	–
	Can	Emergency measures planning requirements		Licence quotas for water allocations/ permits	+	Flood Zone building restrictions	–
		GHG reductions (industry and sector specific)	–			Emergency measures planning requirements	+
	LA	Glacier Preservation Law (Ch8Mz)	+	Water rationing (Truro) Zero Net Land	–	–	
		Fines for illegal water extraction (Ch7Co)	–	Degradation target of UNCCD (Ch8Mz)	–		
		Reversion of unused water allocations (Ch7Co)	–	Ecological flow restriction (Ch7Co)	–		
		Fee for non-use (Ch8Mz)	–				
		Inherence water instrument	–				
		Energy efficiency	–				
		Public participation in environment (Ch8Mz)	+				
Missing		Pervasive GHG reduction, Berlin Rules on Water Resources, Human right to water and sanitation, Climate change1.5–2 degree limit, Climate change lawsuits, Right to be free from climate change damage, Fines for illegal drainage					

(continued)

Table 9.3 (continued)

Econ	**All**	–	+	Crop insurance	++	Emergency relief	+
				Loan instruments	+	Agri-income stability programs	+
				Agri-income stability programs	+	Loan instruments	+
	Can	Tradable rights to water (Ch6Al)		Ag producer water infrastructure grants (FRWIP)	+	Disaster assistance	+
		Water tariffs	+	Agri-innovate fund to provide new technology and global market competitiveness	n.d	Disaster loan (Ch6Al) Ag producer water infrastructure grants	+
		Carbon offset market (Ch6Al)	–	Grants irrigation efficiency (Ch6Al)	+		
	LA	Tradable rights to water (Ch7Co)	–	Micro insurance GEF projects (Ch8Mz)	–	–	
		Water tariffs (ch8Mz)	–	Provision of supply of inputs as relief measures	–		
		Export measures encouraging virtual	+	Fee for non-use of water (Ch7Co)	–		
		water export (Ch7 Co)	–	Drinking water subsidy for poor	–		
		Carbon offset markets (Ch8Mz) FAO TeleFund and Trsut Fund (Ch8Mz)	n.d				
Missing		Payments for ecosystem services, conservation tenders, environmental taxes, bonds, royalties, tax rebates, conservation auctions					
		Direct programme spending on research on climate change mitigation and adaptation					
		Subsidies on products or practices, loans, equity, bonds, crowd-financing and grants, Climate Impact assessments, Adaptation Fund, Development market place, Strategic climate fund, Emissions trading and transaction log (UNFCCCC)					
		Creation on non-farm employment opportunities					
		Payment for ecological services (shelter belts)					
		Flood insurance, Hazards of place indicators of vulnerability, Catastrophic bonds					

(continued)

Table 9.3 (continued)

Suas	**All**	Drinking water quality reports and alerts	+	Provision of information to public	+	Provision of information to public	+
		Provision of information to public	+			Flood alerts	+
		Agri-Environmental programs	+				
		UNFCCC State communications	+				
		Sustainable	+				
		Development goals	+		+		
	Can	FSP for environmental best practices		Drought prediction and alerts		Hyogo Platform for DRR	+
				Drinking water quality		Water security plan	+
		Climate change plan (Ch6Al)	–	reports and alerts		FSP	+
						Right to humanitarian aid	+
						Hyogo Platform for DRR	-
						National Disaster Mitigation Strategy	-
						Red Cross Guidelines for humanitarian aid	n.d
	LA	National climate change plan	–	GEF and CDM projects (Ch8Mz)	–	–	
		Adaptation plan (Ch7Co)		Tariff rebate on building wells (Ch8Mz)	–		
		Glacier Protection policy (Ch7Co)	–	INDAP outreach (Ch7Co)			
		National strategy integrated water management (Ch7Co)_	–				
Missing		Climate change forums, Measures on climate change and environmental awareness and responsibility					
		DRR tools, indicators, best practices to build resilience					
		Government demonstration through practices of procurement, building infrastructure, and processes of environmental stewardship/climate change mitigation					
		Persuasion for water demand management					
		UN Watercourses Convention, UNECE Water Convention					

(continued)

Table 9.3 (continued)

Mgt	**All**	Water attaching to land and managed by owners as common property	+	Hyogo Framework for Action	+	Cooperation agreements for disaster recovery	
		Local water governance (irrigation)	+	International Standards	+	Hyogo Framework for Action strategic goals and seven targets	+
				Risk management tools		International Standards	
						Risk management	+
						World Humanitarian Summit, ISO rules	
	Can	Source water protection plans (Ch5Sk)	+	Insolvency	+	Insolvency	+
		Boundary Water Treaty	+	Drought strategy (Ch6Al)	+	Flood or disaster plan	+
		Integrated water plan (Ch6Al)	–	Integrated land use management and planning	–	Humanitarian aid of ngos	+
				Publlic participation in water strategy, planning (Ch6Al)	–		
				Local watershed plans	+		
	LA	Water registry	–	National irrigation policy (Ch7Co)	+	Flood zone infrastructure plan (Ch7Co)	–
				Public participation (Ch8Mz)	+		
Missing		Long term water management plans on integrated basis, Proactive community planning for water shortages, Demand management of water, Integrated water resource management					
		Community disaster planning for resilience Indicators- Hyogo Framework-Hazards of Place DRR tools, Long term counselling support services post flood disaster, Inclusive participatory development					

All All countries, *Can* Canada, *LA* Latin American countries, *Sk* Saskatchewan, *Al* Alberta, *Co* Coquimbo, Chile, *Mz* Mendoza, Argentina
++ Effective; + Moderately effective; _– Ineffective; n.d Not determined

9.5 Impact on Actors – Effectiveness at Achieving Mandate

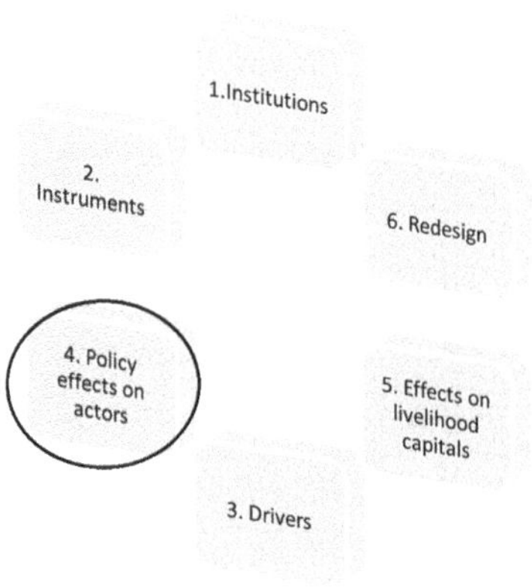

The effectiveness of the instruments in achieving their mandate in relation to the rural agricultural producers, given the drivers, in the case studies (see Sect. 3.3.5) is compiled by instrument category.

As a group, **regulatory** instruments were particularly ineffective, mostly due to lack of resources and the political will to enforce. Illegal water extraction, water reduction quotas, water quality measures, waste water treatment and drainage permits all require significant (and seemingly unavailable – given the driver of government austerity) personnel and policing to ensure compliance. Reversion of unused water rights was also problematic, as water rights holders would employ numerous strategies to avoid loss of water rights. The DCs' practice of proportional water reductions to meet drought conditions was also found not to be effective. These practices solved the immediate problem of shortage, but after having implemented them for several years consecutively, interviewees realized that the *turno* instrument didn't allow innovative adaptive practices that might include changing crop location, sharing water with others or transferring water (permanently or temporarily).

Economic instruments were somewhat effective in relation to adaptation decisions minimizing exposure to d&f, but not accessible for the very small and poor producers, thereby exacerbating inequality. In Canada small producers find them unaffordable; in Chile and Argentina the small producers don't qualify due to such criteria as arrears of taxes. In Chile, it is contested if the water market is efficient, and many individuals and communities were without water rights or access (Reyes et al. 2009). Any court proceeding to substantiate water rights or obtain environmental damages are ineffective as they are costly and lengthy (*ibid.*).

Suasive instruments were particularly effective in providing information and persuading. Providing information on drinking water quality, imminent d&f, allowed actors to make informed proactive decisions. However, the Glacier

Protection Policy (Chile) and Glacier Protection Law (Argentina) were thought ineffective at addressing the root causes (GHG mitigation) of glacier loss.[1]

Managerial instruments are particularly effective in relation to d&f, if done properly on the appropriate scale. D&f events must be proactively planned for in order to reduce their magnitude. Management instruments in relation to climate and water were not as effective. Most areas (with the exception of Saskatchewan) had plans for climate change and adaptation, but little concrete action or impact on climate mitigation (reduction of GHGs) or comprehensive adaptation. Water management instruments such as source water planning and IWRM had limited effectiveness because the plans were unenforceable and unimplemented and development decisions occurred in a separate forum. Alberta had too many plans and consultations, and not enough unification between them. Although the management of water by its owners was somewhat effective through irrigation districts, juntas and inspectorates, those without formal water interests (in DCs) were unable to participate in decisions (the *Campesinas).*

This research found that it isn't the caregory or type of instrument that is effective over the next, but rather the suite of instruments addressing a particular policy problem. For adaptive governance, the 'suite' of policy instruments that adequately respond to a problem are the most important determinant of effectiveness in relation to mandate and achieving ultimately adaptive governance. The multi-level institutional analysis (steps three to five) (see Sect. 3.3) provided for an analysis of a combination, or suite of instruments responding to a particular aspect of d&f. There were three combinations that were particularly noteworthy:

1. *Property interests in water.*

In relation to drought, the water property interest suite influences the equitable allocation of water. In the case study of Alberta, the existence of a regulatory license as well as a limited ability to transfer water interests separate from land contributed to economic capital as well as social capital during the 2001 drought (see Sects. 6.5.2 and 6.9.2). A diversity of instruments on water property can enhance resilience and capitals more than singular instruments in each study region (see Table 9.4).

The property interest in water was optimal in Alberta. Here, a strong institutional system supports government ownership and licensing of water, setting conditions of duration, time, volume, etc.. A limited market enables the transfer of water interests in certain prescribed circumstances creating efficiencies. This system allowed responsive, timely adjustments to be made during the 2001–2002 drought in Alberta (see Sect. 6.9.2). The exclusive existence of a market to govern water allocation decisions, as in Chile, has allowed vast investments in agricultural production. However, this has impacted equity and human capital (health) as the property interests in water overshadows other instruments, specifically the human right to water.

[1] In Mendoza, Argentina this law is effective in achieving its unstated purpose of keeping mining out of Mendoza, and for this reason an example of triple loop learning. Economic growth in mining was rejected because of the damage it would cause the environment and constraints imposed on the water resource.

Table 9.4 The water property interest by study areas

Study area/ principle	Saskatchewan	Alberta	Chile	Argentina
Governing water principle	Common property	Most beneficial use	Public good, right of access is private property/marketable commodity	Public good
Allocation of water rights	Licensed interests allocated by government agency with conditions	Licensed interests with some ability to transfer in certain circumstances	Initially by government; then market	Inherent with land
Water Priorities	No statutory priority scheme	Statutorily legislated model with some water trading	None	Human use, irrigation, industry, then fishing and plant ponds
Dispute mechanism	Water Appeals Board	Government minister, then court litigation	Court of law and local Juntas de Vigilancia	Tomero, Inspector, then Appeals Council
Pricing	Set by municipal water supplier	Regulated by Utilities Commission	One off fee for access right, then service costs; market sets price on transfer	Regulated by EPAS
Water property interest				
Regulatory	Regulatory system overseen by Saskatchewan Water Security Agency	Regulatory system overseen by Environment and Sustainable Resource Development department	Absent after initial allocation	Regulatory system overseen by DGI
Market	Absent	Water market established by Chilean constitution	Limited transfer provided by legislation	Absent
Management	Local watershed councils develop source protection plans	Local watershed councils develop source protection plans	Basin organizations and water tables exist to integrate management	Water is managed by DGI, General Users Assembly, Riverbed Inspectorates
Implications of diversity	Some provision for diversity in license decision making	Diversity provided for in producer, actor and license decision making	Very little diversity as decisions predominantly made by water licencees	Some diversity in decision making, but little in water re-allocation

Table 9.5 Financial instruments of agricultural producers

Study area/principle	Saskatchewan	Alberta	Chile	Argentina
Farm Income Stability Programs	Growing Forward	Growing Forward	No	No
Farm water infrastructure programs	FRWIP	FRWIP	Irrigation programs	
Agricultural loans	Yes	Yes	Yes	Yes
Crop Insurance	Yes	Yes	Yes	Yes
Specialized programs for small farms	No	No	Yes	Yes
Disaster Assistance	Residential premise	Residential premise	Small Farmers	Small Farmers
Bankruptcy discharge	Yes	Yes	No	No

No No instrument available, *Yes* Yes instrument available

2. *Financial instruments (Economic and Managerial).*

Economic and management instruments help develop farmer capacity to respond to d&f. Financial instruments providing capital, loans and subsidies as well as bankruptcy and restructuring instruments can help in adaptation and livelihood transition.

Instruments missing in the DCs are those supporting producers financially through loans and instruments like bankruptcy and insolvency softening a transition to another livelihood. Disaster assistance by way of tax forgiveness in Argentina (DACC 2014), or Chile's small emergency funds (USD300 per family; Reyes et al. 2009) allows subsistence farming to continue (Table 9.5).

3. *Flood instruments.*

None of the study regions have a robust link between institutions and flood prevention. Flood prevention occurs through flood zone mapping, flood zone building restrictions, and building requirements (Henstra and McBean 2009) but a holistic consideration of all these instruments hasn't occurred anywhere. Intricately linked to these measures are insurance and disaster assistance instruments. Insurance and reinsurance companies will not insure floods in the absence of flood risk reduction measures; hence there was no flood insurance in any case study. In Canada, the state pays disaster assistance to homeowners (but not poorer tenants) who often rebuild in the same flood prone location; in the DCs there is no such assistance.

9.6 Impact of Instruments on Livelihood Capitals

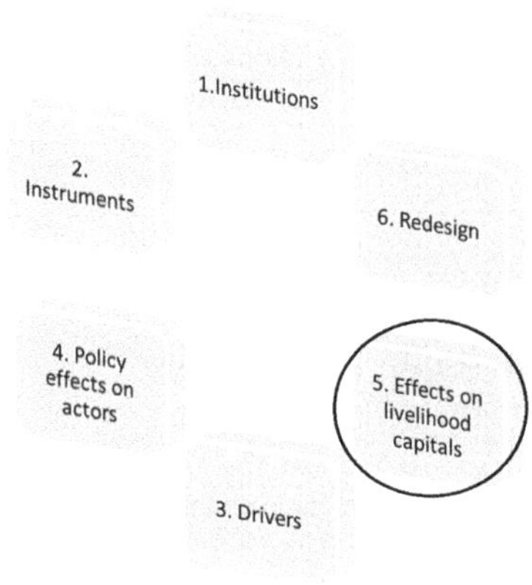

Table 9.6 compares the effects of the instruments on livelihood capitals in each study region.

9.6.1 Human Capital

Direct impacts of d&f related to mental health (stress) on agricultural producers. D&f also impacted the provision of quality drinking water. No instruments addressed the former in the DCs; in Canada, the government contracted counselling and other disaster related services (housing, food, etc.) from a NGO, the Red Cross. In times of water shortage in Chile communities were forced to rely on poor quality water from trucks. This service is not subject to instruments enforcing drinking water quality standards (raising health concerns).

Eduction, in relation to the adoption of adaptive practices by agricultural producers, was found to be directly related to social capital. Agricultural producers learned about new technologies (in relation to irrigation, seeds, equipment etc.), new farming practices, marketing of produce in local to international markets, all through social relationships with other agricultural producers, relations with government

Table 9.6 Impact of instruments on livelihood capitals in study regions

Study area/capitals	Sk	Alta	Chile	Argentina
Human	++	++	–	No data
Social	++	++	+	+
Econ.	+	+	+	–
Techn.	+	+	+++	–
Natural	–	–	–	–

+++ Strong; ++ Medium Strength; + Some Strength; – Less Strength

personnel, cooperatives, and in some case private business companies. Those producers with the most relationships and connections, tended to have the highest informational capital to embrace adaptive practices and build livelihood resilience. Examples appear in Sect. 9.6.2.

9.6.2 Social Capital

Currently, vibrant producer networks exist in Argentina, Alberta, and Saskatchewan for exchanging agricultural information, sharing labour, responding to d&f, and organizing interests in and access to water (Corkal et al. 2016; Warren 2013; M3) without the assistance of government. Similarly, local Chilean community water groups were highly engaged and active in increasing education surrounding water and water practices (Reyes et al. 2009). However, the private water right empowering the development and expansion of irrigated agriculture (wine and produce) rendered the local communal water governance of the Juntas ineffective (Hill 2013; Donoso 2014) reducing trust.[2]

In Canada a style of disaster risk planning of 'incident command' (wherein one department takes charge of the disaster directing the actions of all others) was believed by interviewees to reduce social capital; the planning and response to disasters resided with technocrats, not people (A1). In relation to disaster planning, the missing participatory instrument that includes people and stakeholders has serious implications. It is contrary to the disaster risk reduction triple loop learning at the international level which calls for recognition that effective response includes involvement of people, building of resilient systems, and solving issues of structural social inequality (see Sect. 4.6).

Instruments facilitating irrigation have developed a strong irrigation based social capital in all study regions. Historically in Mendoza and Alberta, and more recently in Chile, government instruments have provided the money to build dams and infrastructure necessary for irrigation and the resulting livelihoods (M2; Reyes et al. 2009; S5; Sauchyn and Kulshreshtha 2010). It is concerning that in Alberta this institutional capital is reducing because of lack of government priority, in Argentina because of lack of capital, and in Chile because of lack of trust and conflict between private water rights holders (Clarvis and Allan 2013). This combined with the driver of increasing government austerity may continue to reduce this capital.

Instruments promoting local watershed planning (all study areas) and local environmental farm stewardship (Alberta and Saskatchewan only) did have a positive impact on social capital as these instruments brought together producers and communities in forums designed to advance community water and environment interests (Davidson 2010; Corkal et al. 2016; Reyes et al. 2009; M5). This counters the

[2] Statistics corroborate these observations. HDI (2014) ranks trust in people in 2011 as a score of Chile 15, Argentina 23 and Canada 42; satisfaction with community is ranked in Chile at 78.4, Argentina 89, and 91.7 for Canada.

loss of social capital from increasing urbanization and migration from the rural sector. In each study region loss of social capital and youth was a concern challenging the viability of schools, churches, sports teams, etc. in rural communities.

9.6.3 Economic Capital

All study regions had similar economic instruments: crop insurance, emergency flood or drought relief, and infrastructure upgrade programmes, which all favourably impacted economic capital. However, two trends in economic capital were noted. First, small producers were often unable to access economic instruments. In Alberta and Saskatchewan the reason cited was affordability, whereas in Chile and Argentina reasons often related to not fulfilling legal requirements e.g. paid up taxes (Reyes et al. 2009; M5).

Second, all case studies revealed an emerging large agricultural producer model of production which was not a policy objective of any instrument, but the result of a combination of the unaddressed drivers and suite of instruments. In Saskatchewan and Alberta large dryland agri-business producers access instruments, increase farm acreage, and diversify risk across geographic locations, utilizing cutting edge technologies (C11). In Chile large, capital-intensive viticulture operations also purchase water rights, access technology and markets (Hadarits et al. 2016; Valdez-Pineda et al. 2014) and in Argentina large viticulturists access groundwater and diversify operations by both producing grapes and making wine (M2). The driver of increasing inequality is a lived reality of the agricultural producer and accentuated in the DCs by barriers to trade for small and medium sized producers (Montaña et al. forthcoming; OECD 2012; C11).

Counteracting this are (1) informal social water practices (and in Argentina the formal inherence water instrument) that take precedence at the local level over laws and rules surrounding water governance; and (2) the built infrastructure of irrigation which allows producers to adapt. In Mendoza local irrigated agricultural producers protect their land and water interests from forfeiture but *Campesinos* are marginalized and human rights to water go unrecognized (Montaña et al. 2005). *Campesinos* suffer the same fate in Chile and the water market allows large mining and energy companies to access water protected by their private property interest and prevent water arriving in the local community.

9.6.4 Technological Capital

The driver of growing inequality is also reflected in access to technological capital. Large powerful producers had the best access to innovative technology in all study regions. In dryland farming in Canada, technology was the biggest area of innovation and learning found in the case study, but out of reach of small producers.

The Chilean Irrigation Development plan in combination with the water market have assisted in the expansion of its export viniculture and horticulture industry with most producers in the Elqui River basin using modern irrigation technology. In contrast, without the water market, and with the centuries old practice of the water right inherent to the land, only 6% of Mendocina's producers use sophisticated irrigation technology (Montaña and Boninsegna 2016).

D&f significantly impacted communities and technological infrastructure. Droughts strained drinking water and sanitation systems; floods destroyed drinking water and sanitation systems, roads, bridges etc. Canada had an instrument to respond – i.e. disaster assistance paid by the state, but the DCs did not. One further deficiency in all case studies was noted – few people link water retention infrastructure built to address drought with that needed to prevent flood (C5).

9.6.5 Natural Capital

There was a deficit overall of effective instruments supporting **natural capital**. In Canada and Latin America ecosystem services are reducing (FPTGC 2010; Herzog et al. 2011). This is consistent with the unaddressed drivers outlined in Sect. 9.2.

9.7 Comparing Adaptive Governance and Policy Structuring

Adaptive governance and policy structuring (see Sect. 2.4.2) within the case studies (see Sects. 4.5, 5.6, 6.6, 7.6 and 8.6) have similarities in relation to risk cognition and policy framing. The areas shaded by dots in Fig. 9.2 represent areas where evidence of policies and instruments existed within the case studies tackling the unstructured, moderately structured, or structured problems of climate change, d&f.[3]

Consistent with the ontological theory underpinning this research, people are fallible learners. The problems of drought were cognized in all study regions and flood only in case study areas in Canada recently experiencing flood. Climate change was least cognized as a risk in Canada where the drivers of denial and scepticism were present (although in 2016 this is changing with the new federal government). These findings are clearly at odds with the science of the IPCC.

There is much room for improvement of adaptive governance. There was very little, if any consideration of climate change and the embedded problems of d&f in relation to policy instruments in any of the case studies. Each of the problems of d&f and climate change were approached separately. To improve adaptiveness, con-

[3]This figure is an illustration of areas where instruments where discovered. Dots are not representative of specific instruments, nor representative numerically of the numbers of instruments.

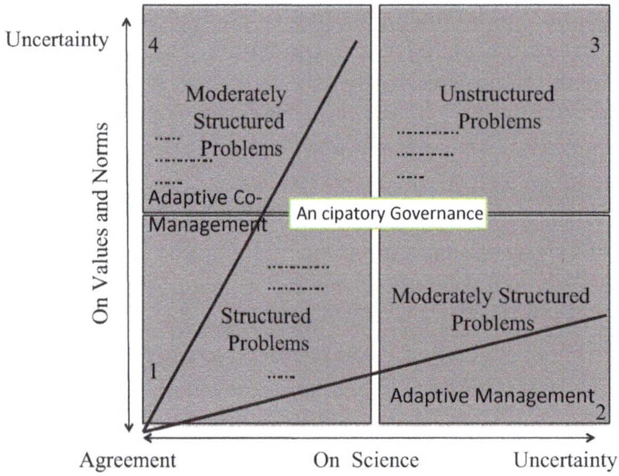

Fig. 9.2 Evidence of adaptive governance in case study areas
DOTS – Depict effective instruments
ACM Adaptive co-management, *Q4ACM* Quadrant 4 moderately structured problem, *ANG* Anticipatory governance, *Q3UP* Quadrant 3 unstructured problem, *Q2MSP* Quadrant 2 moderately structured problem, *AM* Adaptive management, *Q1SP* Quadrant 1 structured problem

sideration of the environment and ecosystem services should be done holistically such that climate change, d&f are considered together.

The theoretical framework of adaptive governance developed in this book has not yet been considered and applied holisitically in the policy realms of the case studies. Figure 9.2 portrays a consistent pattern of not considering adaptive governance and the inclusive components of adaptive management, co-management and anticipatory governance. In the case studies unstructured Q3 problems were not consistently tackled, nor were Q2 adaptive management techniques used to address scientific uncertainty or adaptive co-management and anticipatory governance techniques used to address normative uncertainty. Lapsed strategies and ineffective instruments predominantly occupy the Q3 quadrant and only sporadic incidents of anticipatory governance.

The link between climate change, d&f as a public cognized risk and the defecit of international instruments being implemented at the local community level is aptly illustrated in relation to the 'disaster' problem of d&f. In respect of disaster institutions, Canada, Chile, and Argentina participate in the Hyogo Platform and have filed recent progress reports (Canada 2014; Chile 2013; Argentina 2013). While key policy stakeholders were aware of this instrument, there is little evidence of the influence of the international learnings surrounding DRR at the local level. Triple loop learning in relation to disaster planning (see Sect. 4.6) is not being applied in the local case studies. Canada admits that DRR is largely an emergency management issue and a 'false sense of security' exists in Canada in respect of the perception of risks in the population because Canada has not recently experienced a natural disaster (Canada 2014: 49). Although learnings surrounding public participation in disaster planning, including considerations of climate change, equity and resilient development exist at the international level in relation to DRR, these learnings don't translate into institiutional practices at the Canadian national, provincial or local level.

In contrast, Q1 is well occupied by strategies responding to the structured problems of climate impacts (d&f) and related policy problems of stabilizing agricultural producer income or improving profitability. All case studies displayed a significant disconnect between instruments and policies assisting in climate adaptation, but not framed as such.

9.8 Inferences

This chapter compared the findings in each step of the multi-level case study for three purposes. First, it provides insights carried forward to answer the primary research question in the next chapter: how a theoretical and policy framework can be designed to build capacity of rural agriculturcal producers to respond to d&f. Some of these include: the importance of international economic drivers; the importance of interconnections of formal institutions for adaptive governance; and the implications of growing inequality for adaptation and access to technological, economic, and natural capital.

Second, this chapter determined what policy instruments are effective (see Table 9.3) and which instruments operate within differing contexts and improve agricultural livelihoods. The following instruments were found to positively impact livelihood capitals:

(a) Economic instruments in the IC and DCs are important in addressing economic losses associated with d&f thereby building economic and technological capital. These instruments also have encouraged single loop learning when combined with strong social capital, thus advancing human capital. (Attention should be paid to accessibility of these instruments by small agricultural producers);
(b) Suasive instruments (such as providing meteorological information, drought predictions, flood alerts, environmental best practices etc.) in the IC and DCs were found to be effective at showing the need for adaptation and positively impacting economic and natural capital;
(c) Transition instruments in the IC (bankruptcy, insolvency, basic living exemptions associated with the process such as home quarter protection) allowed agricultural producers to switch livelihoods improving economic capital and human capital (mental health);
(d) Disaster assistance instruments in the IC (the provision of counselling and support services and financial assistance for homeowners) improved human (mental health) capital and economic capital for homeowners;
(e) Managerial instruments (planning for drought and flood, especially in a participatory manner) are important for building social capital. (Extra attention to these is needed in the DCs and increased participatory methods);

(f) Participatory instruments involving many organizations and people and involving deep debate of issues (such as climate change and development in the Mendoza case and source water protection in Saskatchewan) build social capital. Mendoza (even with less state financial support characteristic of DCs) successfully achieved triple loop learning by passing a Glacier Protection Law. The water inherence principle, strong institutional irrigated water, a social capital characteristic of Mendoza's hydrological society, and leadership of civil society, environmentalists, and politicians account for this; the most important variable is a continued use of participation instruments;

(g) It is not the type of instrument (regulatory, economic etc), but the 'suite' of instruments addressing a problem that is important. A flexible suite of instruments surrounding the property interest in water that include regulatory instruments and appropriate market instruments (in the Alberta case) allows adaptation by building economic capital in times of drought.

Third, the case studies illustrate how instruments are framed in formal policy. The cognized risk associated with policy problems of climate change, d&f are three separate realms of policy problems. They are not cognized as embedded, holistic environmental policy problems. Instruments are framed as responding to drought, or flood, or climate change, but not all three. This raises two further insights:

(a) There is room for improvement of adaptive governance systems in respect of d&f in the case studies. This research discovered that adaptive governance is not being fully utilized in the case study areas. Few examples could be found of instruments tackling unstructured Q3 problems with processes of anticipatory governance and adaptive management (hypothesis testing where science is uncertain. Many institutions exist in relation to d&f, especially suasive instruments at the international level (see Sect. 4.3.2). Very few of these international instruments were found operationalized in the case studies. There was a deficit in planning for emergencies of d&f in the developing countries (DCs). Iterative public participation instruments addressing unstructured, complex problems of d&f were missing in all case studies.

(b) Key missing instruments are preventing double and triple loop learning. These include instruments involving participatory policy development and addressing Q3 unstructured problems and Q2 problems with uncertain science and the corresponding anticipatory governance and adaptive management techniques. Although international instruments occupy these quadrants, they are not operationalized in the local case study level. As a result two important triple loop learnings of the international institutions are missing: (1) DRR is not being conducted in a resilient manner that includes the input of people as well as accounting for the mitigation and adaptation to climate change. Instead a realist construction of risk by technocratic experts occurs; (2) Development is not being pursued in a resilient manner by including all people in decisions as well as accounting for necessary climate change mitigation.

References

Argentina (2013) Inform Nacional del Progreso en la Implenentacion del Marco de Accion de Hyogo (2011–2013). Buenas Airies

Bauer C (1997) Bringing global markets down to earth: the political economy of water rights in Chile. 1976–95. World Dev 25(5):639–656

Canada (2014) National progress report on the implementation of the Hyogo Framework for Action (2011–2013). Humanitarian Affairs and Disaster Response Division, Foreign Affairs and International Trade Canada. Ottawa. Available at: www.preventionweb.net/files/29678_can_NationalHFAprogress_2011-13.pdf. Accessed 15 June 2015

Carruthers D (2001) Environmental politics in Chile: legacies of dictatorship and democracy. Third World Z 22(3):343–358

Chile (2013) Informe Nacional del Progreso en la implementacion del Marco de Accion de Hyogo (2011–2013). Santiago

Clarvis MH, Allan A (2013) Adaptive capacity in a Chilean context: a questionable model for Latin America. Environ Sci Policy. Available at: http://dv.org10.1016/j.envxxc.2013.10.014. Accessed 15 June 2015

Corkal D, Morito B, Rojas A (2016) Chapter 11: Values analysis as a decision-support tool to manage vulnerability and adaptation to drought. In: Diaz H, Warren J, Hurlbert M (eds) Vulnerability and adaptation to drought: the Canadian Prairies and South America, ISSN 1919–7144. University of Calgary Press, Calgary, pp 251–278

Davidson D (2010) The social landscape. In: Sauchyn D, Diaz H, Kulshreshtha S (eds) The new normal. Regina, Canadian Plains Research Center

DACC. (Direccion de Agricultura y Contingencias Climaticas) (2014) Website. Available at: http://www.contingencias.mendoza.gov.ar. Accessed 30 Nov 2014

Donoso G (2014) Presentation to workshop. From global to local: sustainable water resource governance for Latin America and the Caribbean. At Universidad Catolica de Chile, Santiago Chile, May 5, 2014

FPTGC (Federal, Provincial and Territorial Governments of Canada) (2010) Canadian biodiversity: ecosystem status and trends 2010. Canadian Councils of Resource Min-isters, Ottawa, pp vi + 142

Galaz V (2003) Privatizing the commons, natural resources, equity and the Chilean water market. FLASCO, Santiago

Hadarits M, Santibanez P, Pittman J (2016) Chapter 13: Drought risks and opportunities in the Chilean grape and wine industry: a case study of the Maule region. In: Diaz H, Warren J, Hurlbert M (eds) Vulnerability and Adaptation to Drought: The Canadian Prairies and South America, ISSN 1919–7144. University of Calgary Press, Calgary, pp 303–326

Henstra D, McBean G (2009) Background report. Climate change and extreme weather: designing adaptation policy. Zurich Canada, BC Ministry of Environment, AMEC, British Columbia

Herzog SK, Martinez R, Jorgensen PM, Tiessen H (eds) (2011) Climate change and biodiversity in the Tropical Andes. Inter-American Institute for Global Change Research (IAI) and Scientific Committee on Problems of the Enviornment. Available at: www.iai.int/wp-content/uploads/2014/06/book.pdf. Accessed 16 June 2015

Hill M (2013) Climate change and water governance. adaptive capacity in Chile and Switzerland. Springer, Dordrecht/Heidelberg/New York/London

HDI (2014) Human development index. Available at: http://hdr.undp.org/en/statistics/hdi. Accessed 15 June 2015

IPCC (2014) Summary for policymakers. In: Edenhofer O, Pichs-Madruga R, Sokona Y, Farahani E, Kadner S, Seyboth K, Adler A, Baum I, Brunner S, Eickemeier P, Kriemann B, Savolainen J, Schlömer S, von Stechow C, Zwickel T, Minx JC (eds) Climate change 2014, mitigation of climate change. Contribution of working group III to the fifth assessment report of the intergovernmental panel on climate change. Cambridge University Press, Cambridge/New York

Montaña E, Boninsegna JA (2016) Chapter 14: Drought in the oases of Central Western Argentina. In: Diaz H, Warren J, Hurlbert M (eds) Vulnerability and adaptation to drought: the Canadian Prairies and South America. University of Calgary Press, Calgary, pp 327–348, ISSN 1919-7144

Montaña E, Diaz H, Hurlbert M (forthcoming) Development, local livelihoods and vulnerabilities to climate change in the Andes. Reg Environ Chang

Montaña E, Torres LM, ElM A, Torres E, Pastor G (2005) Los espacios invisibles. Subordinacion marginalidad y exculsion de los territories no irrigados en las tierras secas de Mendoza, Argentina. Region y Sociedad 17(32)

OECD (2012) OECD environmental outlook to 2050: the consequences of inaction. OECD Publishing, Paris. Available at: http://dx.doi.org/10.177/9789264122246-en. Accessed 15 June 2015

Reyes B, Salas S, Schwartz E, Espinoza E (2009) Chile governance assessment final report. Available at http://www.parc.ca/mcri/pdfs/papers/gov03.pdf. Accessed 15 June 2015

Sauchyn D, Kulshreshtha S (2010) Chapter 17: Agricultural adaptation through irrigation. In: Diaz H, Hurlbert M, Warren J (eds) Vulnerability and adaptation to drought in the Canadian Prairies. University of Calgary Press, Calgary

Valdez-Pineda R, Pizarro R, Garcia-Chevesich P, Valdes JB, Olivares C, Vera M, Balocchi F, Perez F, Vallejos C, Fuentes R, Abarza A Helwig B (2014) Water governance in Chile: availabiity, management and climate change. J Hydrol. http://dx/doi.org/10.1016/j.hydrol.2014.04.016. Accessed 15 July 2015

Warren J (2013) Rural water governance in the Saskatchewan portion of the Palliser triangle: an assessment of the applicability of the predominant paradigms, A thesis submitted to the Faculty of Graduate Studies and Research In partial Fulfillment of the Requirements for the Degree of Doctor of Philosophy, 2013. University of Regina, Regina

Chapter 10
Conclusion

10.1 Recalling the Research Question

Agricultural producers are essential for feeding the world's growing population. They generate employment, income, and savings; contribute to health and prosperity, cultural identity, farm tourism, and household food security, and represent a significant sector in DCs (WWAP 2015). The impacts of climate change are anticipated to worsen in the future and may lead to increasing extreme events of d&f. This, in turn, will increase the vulnerability of rural agricultural producers. The cost and destruction associated with d&f draws attention to the need to plan for and respond to d&f. The institutional governance system wherein this occurs (the water, emergency, climate change and adaptation system) is an important component of adaptive capacity (Termeer et al. 2013) impacting the agricultural producers and their communities.

Considerable uncertainty still surrounds how to address the complexity and systemic nature of the problem of climate change and how to improve governance in order to promote adaptive capacity and enhance agricultural livelihood capitals. This chapter answers the research question:

How can a theoretical and policy framework (norms, principles, and instruments (including regulatory, economic, suasive, and managerial)) be designed to build capacity for rural agricultural producers to respond to the increasing likelihood of d&f, defined by uncertainty?

The impact of instruments available to agricultural producers to assist response to disaster – d&f – have been considered in this study within a framework of adaptive governance that has included a focused concentration on livelihoods, risk, and problem framing. The method was a multi-level institutional case study analysis drawing on seven prominent methods for analysing response to environmental problems (see Chap. 3) that has allowed consideration of individual instruments, as well as the combination of instruments available in order to determine (in the

M.A. Hurlbert, *Adaptive Governance of Disaster*, Water Governance - Concepts, Methods, and Practice, DOI 10.1007/978-3-319-57801-9_10

context of drivers) what instruments work, under what conditions, and ultimately improve livelihood capitals of agricultural producers and their communities. The livelihood capital analysis focused on identifying instruments that enhance human, social, economic, technological, and natural capital (Moser 2009).

This chapter assesses the case study findings from a macro perspective, building on the micro and meso analysis of the last chapter. The findings surrounding learning and participation (inclusive development) (see Sect. 2.4.3, step 4) and the ACWs (Gupta et al. 2010, 2015) of each case study will be compared and analysed in order to inform recommendations for redesign. The goal of this research is to build future response capacity through retention of rural producer livelihood capitals by the redesign of instruments.

Section 10.2 of this chapter determines how a **policy** framework can be designed to build capacity for rural agricultural producers to respond to the increasing likelihood of d&f, defined by uncertainty. The remaining sections answer the research question in relation to a **theoretical** framework. Section 10.3 outlines the contribution to theory of this research by defining the problem and linking literature on risk, social learning and participation. Section 10.4 revisits the elaborated methodological framework developed (see Chap. 3) and reviews what worked well, what did not work well, and makes recommendation for the future. Section 10.5 concludes with the need for inclusive development achieved through adaptive governance and systemic thinking aimed at achieving learning.

10.2 Designing a Policy Framework to Build Rural Agricultural Producer Capacity

This research developed a theoretical framework by constructing a model of adaptive governance (see Chap. 2) operationalized through a multi-level institutional analysis (see Chap. 3) incorporating key elements of adaptive governance to study institutions and instruments responding to d&f impacting rural agricultural producers' livelihoods, in the context of uncertainty. The case studies (see Chaps. 5, 6, 7 and 8) provided an analysis, followed by a comparative analysis (see Chap. 9), and offered insights on how policy instruments and institutions could be redesigned to build capacity for rural agricultural producers to respond to the increasing likelihood of d&f characterized by uncertainty. This is the final step of the multi-level institutional analysis, comparing and building on the case study insights.

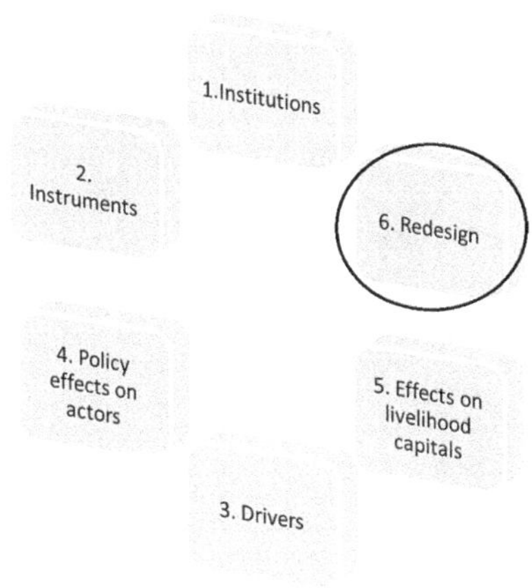

In order to design a policy framework to build capacity of rural agricultural producers three things are required:

1. **A comprehensive institutional consideration of the policy problems of climate change and embedded problems of d&f.** Analysis of the institutional dimensions of adaptive governance using the ACWs, based on the mutli-level institutional analysis, allows for identification of strengths and weaknesses (see Sect. 10.2.1);
2. **A participatory, inclusive decision making approach**. In order to select appropriate effective instruments, disband and revise ineffective instruments, public participation is essential. Not only can past public and stakeholder involvement in policy problem planning be assessed using the split ladder of participation, but future strategies can be formulated (see Sect. 10.2.2);
3. **A concerted focus on social learning** – engaging with uncertainty, exploring policy options through maximizing opportunities for information flow, iteratively exchanging information and evaluating continuous feedback loops (see Sect. 10.2.3).

10.2.1 Comprehensive Institutional Assessment – ACWs

The ACWs below (see Fig. 10.1) represent an assessment of the institutional dimensions of adaptive governance (see Sect. 2.5.3) in each case study. All case studies have room for improvement in the institutional dimensions of adaptive governance. The six major dimensions of the ACWs will be discussed starting with 'variety' and proceeding clockwise. Learning, however, will be discussed in Sect. 10.2.3.

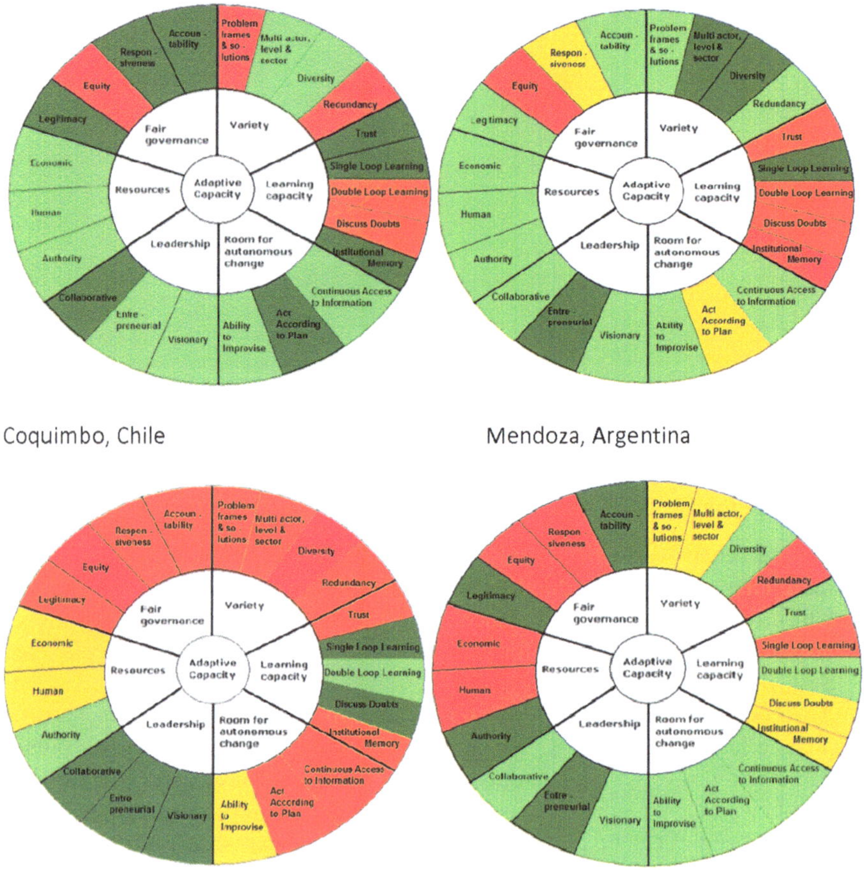

Green	Light Green	Yellow	Light Orange	Red
Very High	High	Medium	Low	Very Low

Fig. 10.1 ACWs of Case Studies

A high score on '**variety**' in Alberta contributes to a high score in most other dimensions. Having a full suite of instruments to address d&f correlates to having a stronger adaptive governance institutional system. Alberta has instruments addressing climate change, d&f and a variety of instruments relating to water governance. Most important was the suite of water property interests (see Sect. 9.5) that include regulatory licenses, economic transfer or market instruments, and instruments facilitating co-management of water. The other three case study areas were missing climate change instruments and a variety of water governance instruments. The private water market right in Chile pervades and in Argentina the water right inherent to land. Improvements to the category of 'variety' can be made by consideration of missing instruments detailed in Table 10.1, as well as effective instruments and instruments that build livelihood capital identified in Chap. 9.

Table 10.1 Unaddressed drivers and missing instruments

Case study	Missing drivers	Missing instruments
All	Climate change, d&f, deteriorating ecosystem services	Pervasive GHG reduction, Berlin Rules on Water Resources, Human right to water and sanitation, Climate change 2 degree limit, Climate change lawsuits, Right to be free from climate change damage, Fines for illegal drainage
	Priority of economy over environment	Payments for ecosystem services, conservation tenders, environmental taxes, bonds, royalties, tax rebates, conservation auctions
	Growing inequality	Direct programme spending on research on climate change mitigation and adaptation
	Demand for energy	Subsidies on products or practices, loans, equity, bonds, crowd-financing and grants, Climate Impact assessments, Adaptation Fund, Development market place, Strategic climate fund, Emissions trading and transaction log (UNFCCCC)
	Government austerity	Creation on non-farm employment opportunities
	Urbanization, population growth, aging producers shortage of farm labour	Payment for ecological services (shelter belts)
		Flood insurance, Hazards of place indicators of vulnerability, Catastrophic bonds
		Climate change forums, Measures on climate change and environmental awareness and responsibility
		DRR tools, indicators, best practices to build resilience
		Government demonstration through practices of procurement, building infrastructure, and processes of environmental stewardship/climate change mitigation
		Persuasion for water demand management
		UN Watercourses Convention, UNECE Water Convention Long term water management plans on integrated basis, Proactive community planning for water shortages, Demand management of water, Integrated water resource management
		Community disaster planning for resilience Indicators-Hyogo Framework- Hazards of Place DRR tools, Long term counselling support services post flood disaster
		Inclusive participatory development
Can	Climate change denial, scepticism	Regulatory water rationing
	Increasing size of farms	Zero net land degradation
		Irrigation development policy
		Glacier protection
		Drought strategy (Sk)

(continued)

Table 10.1 (continued)

Case study	Missing drivers	Missing instruments
LA	Increasing trade liberalization	Emergency measures planning requirements, Hyogo platform at local level
	Neoliberal market and colonial elitism	Flood provisions
	Private social support system and lack of transparency (Ch)	Disaster financial assistance, loans
	Price/currency fluctuations (Ar)	Water infrastructure grants, innovation grants
		Insolvency
		Drought strategy

All All Countries, *Can* Canada, *LA* Latin America, *Al* Alberta, *Sk* Saskatchewan, *Ch* Coquimbo, Chile, *Ar* Mendoza, Argentina

Not only is it important to have a variety of instruments to address the policy problem of climate change and embedded policy problems of d&f, but the interconnection of the variety of instruments is also important. Without effective climate mitigation instruments, more will be required in relation to climate adaptation, d&f.

A defecit of climate change instruments coincides with lack of cognized risk surrounding climate change and the international economic drivers identified in Sect. 9.2. The World Economic Forum identified the lack of attention of global leaders to the Earth's environmental system, and continued attention on the global economic system as providing the potential to trigger the perfect "global storm" (WEF 2013: 11). International environmental instruments at the international level must be adopted and applied by national and provincial/regional governments at the local municipal level. More instruments are required at the national and provincial levels to assist in the reduction of GHG emissions, especially in the developed country of Canada.

'Room for autonomous change' was highest in Mendoza, Argentina. This is partly due to the fact that no barriers were identified in Mendoza for autonomous change. Strong, interconnected cohorts of agricultural producers interact in irrigation groups and marketing groups exchanging information through social interconnections. Comprehensive irrigation planning through Mendoza's institutions responds to drought, acting according to plan and with some ability to improvise at a local level. The scores are not 'very high' as the principle of inherence could be improved in times of emergency drought declarations by allowing some forms of transfer (Hurlbert and Mussetta 2016) and very large agricultural producers do have access to more information than others.

In Chile, interviewees' perceptions were that the water market detracted from agricultural producers' ability to improvise as energy companies and very large producers own water rights. Large private companies fund research and aquire data, but it isn't publicly shared, resulting in reduced access to information. In Saskatchewan agricultural producers access information through social connections improving their abiity to improvse, plan, and adapt. Local watershed groups conduct source water protection plans and environmental farm plans protecting the environment. This is also the case in Alberta, however, many plans and reports are written after disasters of floods, or public consultations, and are not implemented.

'Leadership' was strong in all the case study areas, but different. In Chile and Argentina strong local water groups and environmental groups exist. In Chile the collaborative leadership of local water groups and agricultural producers was noteworthy. The water market created a strong entrepreneurial focus and large agricultural producers and government enact visionary irrigation planning. Mendoza's oasis was similarly created with such visionary leadership as well as the building of the dam to improve adaptive capacity of irrigators. The depth and breadth of the irrigated viticulture and produce economy in Mendoza is highly entrepreneurial. These leadership characteristics also exist in Alberta in relation to irrigators. In Alberta and Saskatchewan dryland producers also exhibit entrepreneurial and visionary leadership in adapting to d&f through new practices and institutions. Strong collaborative local water governance protects natural capital in Saskatchewan. In Canada there are strong small and large agricultural producers.

More leadership is required in Canada from the government for focusing on the discussion of doubts (scepticism) in relation to climate change and modifying current instruments to account for climate change (S1, C4, C7). The Canadian federal government's significant funding cuts to science and prevention of scientists releasing results and speaking to the media (C1) stymied progress.

The main challenge in relation to **'resources'** related to unaddressed international economic drivers outlined in Sect. 9.2 and Table 10.1. Agricultural producers in all of the case study areas identified the challenge of international markets as a key vulnerability of their livelihood. This was especially the case in Argentina where agricultural producers are explosed to international market prices for their produce as well as inflation and interest rate exposure due to the Argentinian government's financial policies.

The interconnection of different levels of government (national to regional/provincial and local) is particularly important in relation to the scoring of resources. The power differentials discussed in Sect. 9.3 impacted the allocation of state resources such that the strong central Chilean government in Santiago resulted in an impoverished regional and local government in Coquimbo.

The challenge for government resources within the institutional system stymied the strong social and human capital that existed at the local level in Argentina and Chile. Local water groups in Chile and irrigators in Mendoza lacked resources to leverage their plans and activities for meaningful change.

All four countries had legitimate government in respect of **'fair governance**.' Coquimbo, Chile's lower rankings reflect the distant relationship from the strong central government in Santiago and interviewees' distrust in the accountability and responsiveness of this far away government. A similar dynamic existed in Mendoza, Argentina with respect to the responsiveness of the central Argentinian government.

Issues of equity exist in all case study regions in relation to fair governance. All assessments of economic capital (see Sects. 5.9.3, 6.9.3, 7.9.3 and 8.9.3) show a growing spread between small/subsistence producers and large/multinational producers; this limits the access of small and medium sized producers to other capitals – technology (to improve effecieincy), natural (groundwater or location closer to the water source), and human (information on adaptation). Increasing trade liberalization (and in the DCs neo-colonial practices) coupled with the missing participatory instruments of inclusive development, especially in relation to Q3 unstructured problems of climate change, d&f exacerbate inequity. Drivers of government austerity and growing income inequality further exacerbate equity.

Local water practices countered inequality in all case studies. However, there is risk that these instruments will be lost. In Chile, these informal practices are not contained within the constitutional water market laws and require recognition by a formal instrument to counter the water market. In Argentina the formal inherence water instrument is being avoided through groundwater licensing. In Canada, the mechanism is recognized as a formal instrument (water transfer) that hasn't been replicated in a decade since the last drought.

In considering institutional redesign to remedy low rankings, the adaptive governance policy framework assesses drivers impacting producer vulnerability and targets these where appropriate. Missing instruments identified in Table 10.1 (appropriate for the given context) might be adopted and effective instruments, especially those increasing agricultural producer livelihoods. Ineffective instruments and practices could be revised or disbanded.

Table 10.1 lists all drivers identified in all countries, as well as drivers specific to each country. In addressing drivers there must be attention paid to unintended consequences. In Argentina, education subsidies were encouraging rural to urban migration as *Campesinos* could attain education for jobs within the Mendoza oasis or city. Discouraging education to address the driver of urban migration might do more harm than good. Similarly, high-energy costs were being addressed in Argentina with subsidized energy. This, however, was encouraging the pumping of groundwater, depleting aquifers and natural capital. More research is required in relation to these unaddressed drivers and their relationship to livelihood capitals.

An internationally binding legal agreement on reducing GHG emissions is required with enforceable commitments from nation states to address the driver of climate change, d&f. The Paris Agreement of 2015 does not go far enough (Gupta 2016). Although the case study areas had many historic strategies surrounding climate change adaptation and mitigation referencing the UNFCCC, there were few effective instruments and not enough variety and redundancy of instruments to ensure success. Although the international level wasn't fully assessed, the analysis in Chap. 4 concluded that in order to take advantage of the multitude of international instruments and learnings, a more direct and efficient connection between the local and global is required. This interconnection should include all people in order to advance key international triple loop learning: forging resilient development pathways that reduce GHG emissions, considering the nexus interconnections of water energy and food, and planing, responding, and recovering from d&f in a holistic manner considering issues of climate change, inequity and poverty.

The Hyogo Framework was recognized by emergency planners in three of the case studies (Coquimbo, Saskatchewan, and Alberta), however, the triple loop learning associated with this international instrument surrounding DRR and resilience (see Chap. 4) was not being implemented at the case study level. Many other important international suasive instruments also exist (see Sect. 4.4) and were not applied in the case study areas. More bridging and linking of international institutions and instruments to the case study level is required.

Especially important is recognition of the right to water and humanitarian aid. Coquimbo had the most precarious right to water for human consumption that interviewees attributed to the exclusivity of the water market. Mendoza had the richest system of water co-management, but only for water rights holders and only in relation to decisions pertaining to the administration of the irrigation system. This situation needs to be remedied; each case study area should assess which missing instruments are appropriate for their situation in conjunction with participatory, inclusive decision making described next.

10.2.2 Participatory, Inclusive Decision Making

The split ladder of participation model allows for both an assessment of participation, policy framing and learning in the past as well as provides a tool for developing strategies surrounding policy problems for the future. First, the mechanisms used in each case study for public and stakeholder participation and climate change, d&f are recounted. There are significant differences in participation and policy problem framing in each case study (see Sect. 2.4.3).

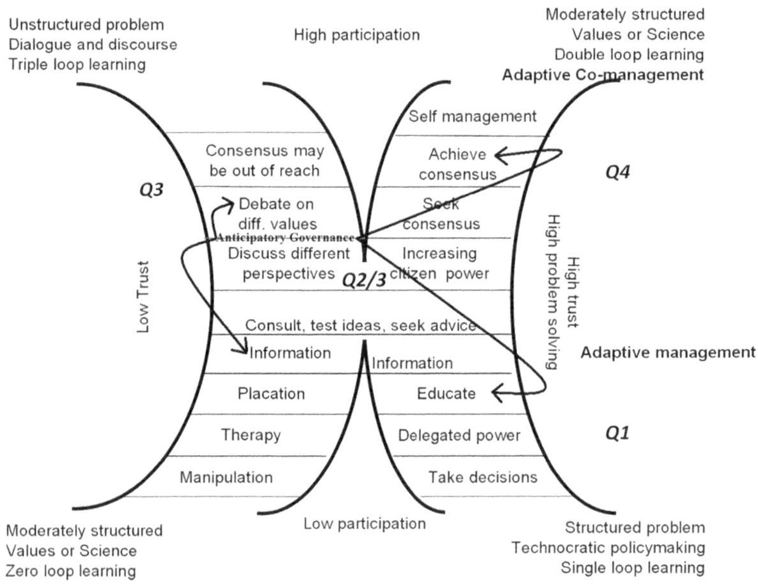

No sustained iterative problems involving public consultation were found to occupy Q3 addressing wicked, unstructured problems in the case study areas.[1] In Canada a federal position on climate change mitigation and adaptation oscillates in and out of favour with no real advancement past a few regulatory measures relating to vehicles and coal fired power plants. There has been no public engagement on climate change policy in Saskatchewan for many years. In Canada, local water groups tackle Q3 climate change problems with source water protection planning through public participation. However, these plans are now almost a decade old in Saskatchewan and haven't been revisited or revised. These local water groups are also closely associated with local communities, towns, villages, districts, and rural municipalities with many common actors in both groups. These groups have historically been very closely connected with significant community participation and engagement making important contributions to the livelihood and viability of agricultural producers.

Alberta also has a long history of including the public in climate change, water, and development consultations. Consultations were held on Alberta's climate change strategy, water strategy, water strategy renewal, water plan, and integrated land plans, amongst others. However, these consultations have been fragmented, involving different actors, and are not interconnected. Because of this, it is questionable how effective the consultations are, and how much true engagement is occurring in relation to the unstructured problems of climate change, d&f in Q3.

In Chile issues are increasingly responded to with technical solutions and very little citizen participation (Reyes et al. 2009). Q3 consultations occurred in Chile

[1] Note the references to fragmented consultations on one off issues described above in Sect. 9.7.

surrounding water governance, climate change and drought in 2014 after the successive years of emergency drought declaration, but more often in the capital of Santiago, and not the study region of Coquimbo. At these consultations the technical solution of diverting water from areas with plentiful water through a pipeline to water scarce regions often arose. Although a similar system of irrigated water governance as in Argentina exists in Chile, interviewees believed its federal bureaucratic structure and the water market removes most policy problems from Q2, Q3 and Q4 of moderately structured problems to a purview outside Fig. 9.2, within the realm of private actors who own water licences.

In Mendoza there were significant disagreements surrounding norms and values that were publicly debated over the last several years. Many public consultation processes occurred surrounding the adoption of a Glacier Protection Law and more recently integrated land planning. These consultations included many months of community engagement (see Sect. 9.8) and many of the same actors participated in all of the consultations. There was much evidence of engagement of producers within other *fora* including producer cooperatives, government and industry meetings, etc.

All case studies had Q4 adaptive co-management in relation to irrigators tackling moderately structured problems in relation to the real time allocation of water within their irrigation district. Mendoza's water governance structure for example occupies Q4, implemented through riverbed inspectors, a general user assembly, and irrigator cooperatives. These bodies display enduring aspects of adaptive co-management (for those with water rights associated with their land, but with the exclusion of all others). These formal institutions were important actors in public and stakeholder consultation surrounding the policy problems of d&f and should also be considered in future public engagement.

The split ladder of participation offers a strategy for the future. Through the split ladder of participation's public engagement, the appropriate missing instruments could be identified. For example the adoption of a water market instrument in Mendoza would be highly contested (M1) and most likely inappropriate to address the wicked Q3 climate change issue. Modifications of the current rigid inherence water property interest that is causing maladaptation will only be effective if the Mendoza people are involved in the selection and implementation of appropriate instruments (Hurlbert and Mussetta 2016).

The split ladder of participation illustrates the missing instruments addressing Q3, the lack of cognized risk surrounding climate change, the driver in Canada of scepticism and denial, and the missing environmental governance approaches (adaptive management, adaptive co-management, anticipatory governance). This gap in adaptive governance explains the continued driver of climate change and deteriorating ecosystem services. There is much opportunity for improvement in adaptive governance by filling this void. Filling this void potentially provides a base for resilient, inclusive, development – development that doesn't increase GHG emissions.

The split ladder of participation also provides a mechanism by which to tackle unstructured and wicked policy problems such as climate change and by utilizing environmental governance approaches of adaptive management, adaptive co-management and anticipatory governance in relation to different missing instruments. For example, reductions in CO2 emissions in agricultural environmental best management practices could be monitored and documented through iterative adaptive management. Policy instruments responding to flood disasters could engage people for more resilient results both before and after flood events.

10.2.3 Focus on Learning

Both the ACWs and the adaptive governance and split ladder of participation model allow for analysis of social learning. Social learning is an important component of adaptive governance (see Sect. 2.4.3) and participation of people is fundamental in order that adaptive and mitigative practices are adopted. Public participation is required for second loop learning – the questioning of mental models – and ultimately triple loop learning – the change of values and norms needed for a deeper understanding of context and transformation to a different social state.

All study areas have adapted to past d&f through intensive single loop learning and economic instruments that provided funds for new water infrastructure, technology and practices. These instruments have predominantly resided within Q1. Other Q1 examples included technocratic policy making surrounding regulation of drinking water, sanitation, or emergency planning in relation to d&f with low levels of participation. Some policy problems (such as enforcement of water quality standards or carbon taxes) are appropriately managed in Q1. However, adaptive governance and learning require periodic reassessment, with public input, which was not occurring in the case studies. It is disconcerting that the problems of d&f and responding instruments predominantly are located within Q1, managed by technocrats without inclusive participation. The deficit in engaging Q3 problems of climate change with public participation, together with the drivers of climate scepticism and denial (accentuated in the DCs by lack of state resources) prevents the international triple loop learnings (see Sect. 4.6) from being embraced locally and increases the vulnerability of agricultural producers and their communities to climate change.

Examples of policy problems in the bottom left hand corner of the split ladder of participation in Coquimbo, Mendoza and Alberta with no learning (zero loop learning) related to the building of dams. Significant groups of people were not consulted but received one-way information flows resulting in low levels of trust, manipulation, and placation. These are not examples of adaptive governance.

All case studies had historic double loop learning that accompanied development of irrigated agriculture. The movement from dryland agriculture to irrigated agriculture represented a change in assumptions and models underpinning strategies in Q4. Only Chile currently has instruments currently supporting irrigation development through an irrigation strategy, water strategy and the private water market. However, the privatization of water has prevented many people from realizing their human right to water (Reyes et al. 2009) and eroded equity and trust by not benefitting small producers (see Chap. 7).

Although Chile's privatized water market, and irrigation water strategies facilitated double loop learning and irrigation expansion, it cannot be concluded that all three of these instruments, to the exclusion of other instruments, such as the recognition of the human right to water, local water practices, environmental practices and flows, will continue social learning in the future. At this moment, the continued identification of irrigation expansion in Chile, to the detriment of these other instruments cannot continue to be classified as double loop learning. Now, another form of double loop or triple loop learning is required in order to recognize the human right to water and remedy the fact that some people in Chile have no access to clean drinking water. The suite of water property interests must be considered. This case study demonstrates the importance of continuous review of wicked unstructured problems and involvement of all people.

One Q3 problem where triple loop learning was identified is in the Andes outside Mendoza. The local social movement (Popular Water Assembly) was able to effectively prevent mining development at the headwaters of the river. The predominant model of economic growth and expansion was effectively challenged and a deep understanding of the importance in Mendoza of the oasis and viticulture was endorsed. Argentina, where the highest **learning** score occurred in the ACW, had some of the lowest scores in human and economic resources. Higher scores for responsiveness, multi-actor/level/sector, leadership, and diversity of solutions might account for this triple loop learning. To achieve triple loop learning (as in Argentina) leadership of multiple people at several levels was required and continuous revisitng of Q3 issues. Climate change and adaptation are still Q3 issues occasionally entering the foray of public debate in Mendoza. Both civil society and regional government leadership achieved this solution through iterative intensive dialogue.

Some parallel can be seen with the triple loop learning examples at the international level (see Sect. 4.6) resulting from robust dialogue between civil society, NGOs and states. These examples show that the institutional context of governance expanded, the power of decision making diffused, and trust created. These processes realized the benefits of participation in decision-making.

The greatest learning (triple loop) found in Mendoza, Argentina did not coincide with a high ranking of variety or resources as anticipated (found in the IC of Canada), but with the high rankings in Mendoza in relation to collaborative, visionary leadership, responsiveness, multi-actor/level/sector, and diversity of solutions. Leadership in this situation occurred through a variety of actors including politicians, civil society, agricultural producers, environmental groups, and academics. Therefore, focusing on leadership at all levels is important for social learning.

Low double loop and triple loop learning reflects: limited variety of 'redundant' instruments with different problem frames across multiple scales and sectors to solve climate change; minimal participatory instruments addressing the unstructured Q3 problems of d&f; lack of adaptive co-management and anticipatory governance approaches; and lack of local operationalization of international instruments applying resilient DRR and inclusive development, while reducing GHGs. In

Canada there were lower rankings for discussing doubts in relation to learning because of the unaddressed drivers of climate change denial and scepticism. This may be one factor preventing double and triple loop learning.

10.3 Contribution to Science

This research advanced science in three ways. First, the model of adaptive governance developed (see Chap. 2) and operationalized in this research (see Chap. 3), advances science by responding to the problems of climate change. These problems include d&f, uncertainty and risk, the systemic nature of the 'wicked' problem, complexity and the contested values in seeking appropriate solutions (see Sect. 1.3.1)), contested science, and the fact that poor governance exacerbates the problem. Second, this research advances science as it improved on the methodology of Young et al. (2005) by including the drivers of vulnerability, linking Young's model with livelihood capitals furthering the adaptive capacity story line, and including elements of risk (including critical realism), policy framing, and the ACWs (see Sect. 10.2). Third, based on this research, recommendations are made for future research (see Sect. 10.4.3) for enhancing adaptive capacity and ensuring inclusive development (see Sect. 10.5).

10.3.1 Uncertainty and Risk

Climate change science concludes that the study areas are warming and more events of d&f will occur (see Sect. 1.3) but there is uncertainty surrounding when and where events will occur, their intensity and duration. This uncertainty is magnified by people's understanding of the risk, mediated and experienced based on geography and culture (Preston et al. 2009); it is socially constructed (see Sect. 2.4.1).

An innovation of this research has been the addition of critical realism and the social construction of risk to both adaptive governance and the methodology of Young et al. (2005). This research has been premised on critical realism and the understanding that natural and social realities exist, but they are also socially constructed (see Sect. 1.8). What proactive actions agricultural producers take in relation to extreme events of d&f, and governments and policy makers plan for, is based on how they construct the risk, and judge the magnitude and impact of d&f. This exercise is both an individual as well as a political act (Leiserowitz 2007; Fischhoff et al. 1984).[2] The performance of this act of construction is not always accurate as information is often incomplete, learning is not always accurate, and perceptions of benefits, costs and reciprocity of relationship differ (Ostrom 1998, 2010).

[2] Often the perceptions of policy makers and people are different (Hurlbert 2013).

This theoretical premise was operationalized by a content analysis of institutions, organizations, and policies responding to climate change, d&f in relation to agricultural producers which revealed if and how these risks are constructed (see Sect. 2.4.1). (A risk must be constructed as a problem in order to be addressed by an institution and a policy instrument). Figures 5.2, 6.2, 7.2, and 8.2 illustrate this risk construction and Sect. 9.7 and Fig. 9.2 summarize it.

10.3.2 Systemic Nature of the Problem

Climate change, d&f are systemic problems impacting many social structures including GDP, insurance, agricultural profitability, and food security (see Sect. 1.2). Adaptive governance provides a systemic solution to the problem. This book developed a model of adaptive governance incorporating the environmental decision making approaches of adaptive management (hypothesis testing), adaptive co-management (distributed decision making), and anticipatory governance (scenario planning) (see Sect. 2.3) and enhancing the literature of adaptive governance with that of risk (see Sect. 2.4.1), problem structuring (see Sect. 2.4.2), and participation, inclusive development, and social learning (see Sect. 2.4.3). Further, the addition of triple loop learning addresses the systemic nature of the problem. (Triple loop learning occurs when values and norms underpinning worldviews and assumptions are changed to arrive at a deeper understanding of context, power dynamics and values influencing management of natural resources that underpin worldviews (Pahl-Wostl 2009; Gupta 2014)).

This research found that the rich literature surrounding these elements of adaptive governance remains largely academic, with little application in practice (although there are positive developments summarized in Sect. 10.2). Very few instances of adaptive management, adaptive co-management, and anticipatory governance, double loop and triple loop learning were discovered in the case studies. Also, very seldom were the policy problems of climate change, d&f linked together, but instead they were dealt with in an isolated, realist risk constructed manner. This knowledge is valuable as it helps explains why so little progress has been made in addressing climate change, d&f.

10.3.3 Complexity (Contested Values and Science)

The problem of climate change, d&f is complex. The uncertainty of science, and systemic impacts exacerbate the uncertainties surrounding optimal adaptation policy (Haque and Burton 2005). The adaptive governance model developed in this book addresses this complexity through the methodological goals of reducing risk, increasing livelihood capitals of agricultural producers, and enhancing the adaptive capacity of institutions (see Sect. 2.5).

Resolving pathways of adaptation is complex as the underpinning values that inform choices and the science of climate change is contested (see Sect. 1.3). Impacts of climate change are borne unequally as marginalized people and those living in poverty bear a greater proportion of impacts (see Sect. 1.3). The adaptive governance model addresses this complexity with participation and social learning. An improvement on this is made with inclusive development in Sect. 10.5. The adaptive governance and split ladder of participation model (see Fig. 2.4) elaborates on what types of learning, what types of policy problems, what types of environmental decision making approach (adaptive management etc.) is appropriate for tackling climate change and embedded problems of d&f. Decision makers, policy people, and academics can utilize this model.

10.3.4 A Better Theoretical Framework to Address Poor Governance

This research developed a better theoretical framework of adaptive governance. An expanded definition of adaptive governance (including adaptive management, adaptive co-management, anticipatory governance) addressed the shortcomings of adaptive governance (deficient methods, solving conflicts of values and science, over broadness, lack of measurable indicators) by incorporating literature on risk, problem structuring, participation, learning, livelihood capitals, and the institutional dimensions of adaptive governance (which identify practices of good governance) (see Chap. 2). This theoretical model of adaptive governance was operationalized through the multi-level institutional analysis (see Chap. 3). Two new models are developed: Adaptive governance and problem structuring (Fig. 2.3) and adaptive governance and the split ladder of participation (Fig. 2.4).

Poor governance is identified in the model of adaptive governance and split ladder of participation (see Fig. 2.4) when unstructured and moderately structured problems with contested science are inadequately addressed. Low levels of trust result such that consensus is out of reach; when low levels of participation occur, placation, therapy and manipulation result.

This research discovered zero loop learning only in relation to the construction of dams (although in other policy areas such as energy it has also been found to exist; Hurlbert and Gupta 2015). This finding was surprisingly not made in relation to water governance, especially in DCs where there are clear winners and losers in water allocation decisions especially in relation to those without water rights. However, mistrust and manipulation were not found (both elements of zero loop learning (*ibid.*)).

10.4 Contribution to Methodology

This book targeted the need for innovative methodologies (Pahl-Wostl et al. 2013) as it developed a multi-level institutional analysis method in order to interrogate adaptive governance and its institutional focus. The analytical framework of this research built upon seven prominent methods for analysing responses to environmental problems (see Sect. 3.2).

10.4.1 Methodological Strengths

The methodology had several strengths. First, it provided a method of applying and evaluating adaptive governance. Second, it provided clearer outcomes to measure in relation to adaptive governance including: (a) assessment of policy problem structuring of climate change, d&f in relation to risk; (b) ranking of effectiveness of instruments on actors in relation to achievement of stated mandate in the context of drivers; (c) assessment of impact on livelihood capitals of instruments; (d) assessment of learning and participation utilizing the model of adaptive governance and split ladder of participation (Fig. 2.4); and (e) assessment of the ACWs of the case studies in order to inform instrument redesign.

The methodological analysis allowed for identification of how climate change, d&f are being addressed and illustrated gaps of policy problems, instruments, and participation. The adaptive governance and split ladder of participation model developed in this research also offers a methodology of tackling policy problems and understanding when stakeholder participation is likely to work (based on factors of trust, values, uncertainty) and under what conditions it is needed (not being required in all structured policy problems at all times). The model requires policy makers to consider and acknowledge the important issue of learning, both what learning is required, and what learning is desired, when considering the policy problem. Focused consideration of unstructured, complex, and systemic problems of climate change, d&f is facilitated. By adding appropriate public participation to this, it is anticipated that democracy can be deepened.

10.4.2 Methodological Challenges

The challenges of this methodology are that there is still an element of normativity in three processes: the assessment of the effectiveness of instruments, the assessment of the impact of instruments on livelihood capitals, and the ranking of the institutional system of each case study according to the ACW (see 'Limits' Sect. 3.5). (In this research any errors in these assessments are mine, perhaps due to my own normativity and ethnocentrism).

The method also suffers from the implications of generalization. It was very difficult to average some dynamics into one score. For instance, in Chile in respect of 'authority' in relation to resources, the *Juntas* were found to be very ineffective (Hill 2013; Donoso 2014) but the institutional support of the Chilean irrigation plan was laudatory. Further, 'economic resources' were available for large agricultural producers in all study areas, but not for small or marginalized producers.

The analysis of learning was also challenging. Assessing if double loop learning (the questioning of assumptions and mental models underpinning strategies) and triple loop learning (a change in understanding in context or a change in world view) (Argyris 1999; Keen et al. 2005) occurred was based on an assessment if such change had occurred, or if social structures had changed. These conclusions were based on analysis of laws, policies, and instruments, secondary sources, and interviews with key informants. Although previous vulnerability assessments were reviewed wherein the perspectives of agricultural producers were accessed, the agricultural producer perspective was represented by only a small number of key stakeholders interviewed. This was a weakness given the focus of this research on the local agricultural producer and their community.

When building this methodology, I did not anticipate that there would be so many organizations and instruments and that the analysis would be so complicated. This resulted in a disproportionate emphasis on formal institutions. Each of the six steps of the multi-level institutional methodology were carried out separately, and because of the depth and breadth of information in each step, in a disjointed fashion. A lengthy process was undertaken of revisiting data gathered in each step, streamlining data, linking the data gathered in each step, and selecting the most important data by continuously revisiting it.

Lastly, the analysis of drivers and their impact on livelihood capitals in relation to the instruments is sketchy. Multiple sources of information were linked together by the researcher. Key informants interviewed were able to shed only some light, based on their perceptions, on this issue. The conclusions here can be seen as hypotheses for future testing.

10.4.3 Recommendations for the Future

Although the complexity of the methodology was a challenge, the nature of the unstructured problem of climate change requires that a complex methodology be employed. Stirling (2010) argues the multiple methods need to be employed and ambiguity, pluralism and conditionality must be accepted in scientific evidence because of the inherently complex and uncertain phenomena that we are dealing with. Pahl-Wostl (2007, 2009) argues that global environmental change requires a complex system analysis in order to manage transitions over multiple time and space scales.

To address the limitations and shortcomings of this research methodology, the following modifications are proposed. In the future, it is recommended that the

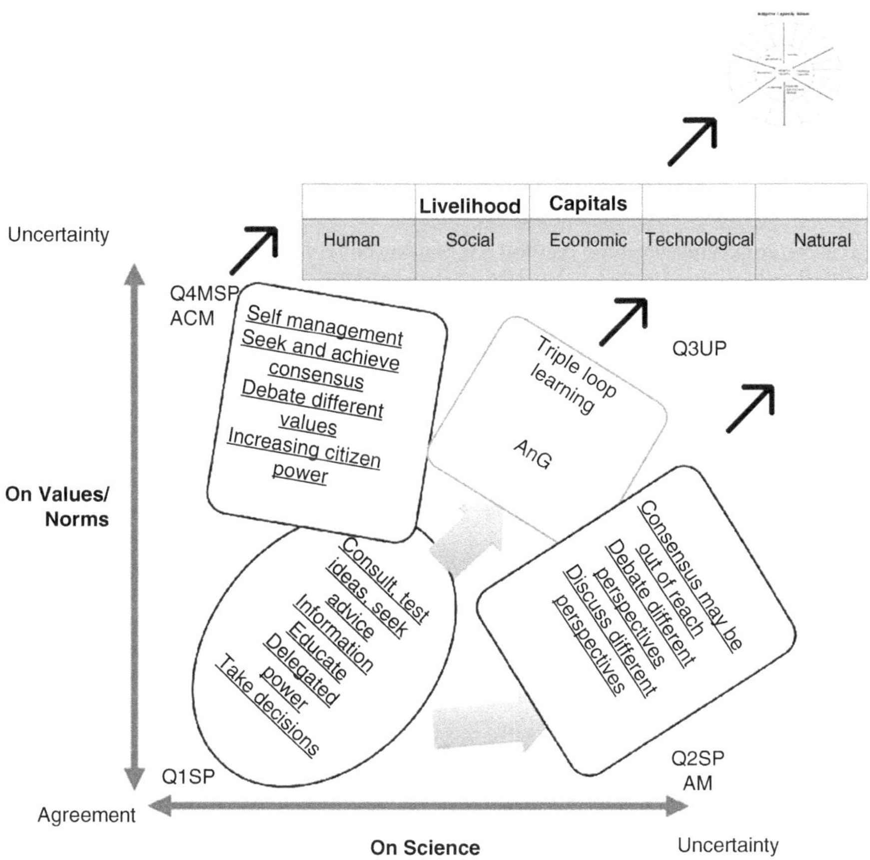

Fig. 10.2 Multi-level institutional methodology: livelihoods and ACW

methods (qualitative semi-structured interviews and content analysis (see Sect. 3.4)) be streamlined, reducing the number of steps (detailed below in Fig. 10.2), and these interviews should be buttressed with a quantitative survey.

Successes

First, the streamlined methodology is easier to conceptualize in the model illustrated in Fig. 10.2. The first stage of the research (currently steps one, two, and three of this research (see Sect. 3.3.1)), would be to identify current drivers, organizations, instruments and policies to populate Fig. 10.2 through content analysis coding for: problem structuring, environmental governance approach, and level of participation, gaps in policy problems, and the utilization of adaptive governance approaches (adaptive management, co-management and anticipatory governance).

Here, I would add another component involving an assessment of the effectiveness of the combination of instruments. Legal pluralism argues that there are often multiple rules applying to the same jurisdiction. In order to understand whether the

Table 10.2 Checking for the impacts of legal pluralism: assessing the quality and intensity of the relations between different policy instruments

Quality/intensity	Weak relations	Strong relations
Contrary	Type 1: indifference	Type 2: competition
Affirmative	Type 3: accommodation	Type 4: mutual support

Source: Bavinck and Gupta (2014)

rules are consistent or not, a method can be formally applied to assess the combination of the instruments applied (see Table 10.2). Sometimes the rules are indifferent to each other and thus there is no relationship; sometimes they actively compete with each other and can result in counterproductive impacts. Sometimes they accommodate each other, and this is often when participatory instruments allow for policy redefinition. Sometimes they mutually support each other. The key challenge is to address rules that actively compete with each other. For example rules on human rights to water and sanitation compete with rules that require that these services are seen as 'economic services' that always need to be paid for (Obani and Gupta 2014).

The second stage of the research would be gathering primary data utilizing three methods:

(a) First, a survey would be conducted that would allow a greater number of participants to contribute to the important themes of effectiveness of instruments, impact on livelihood capitals, presence of social learning, and recommendations for instrument redesign. A large number of participants could increase reliability of data and allow different groups of people (agricultural producers, government personnel, scientists etc.) to be included. The survey would reduce the limits and challenges of this research surrounding normativity and generalization. This information would be analyzed and inform the next step;
(b) Second, semi-structured interviews would be conducted with key policy stakeholders to further explore key themes discovered in the survey. The ACW of the institutional regime would be constructed with the data obtained in these interviews;
(c) Lastly, focus groups would be conducted with people and key policy stakeholders to review information obtained from (a) and (b), and to develop strategies to achieve greater efficiency of instruments, maximize livelihood capitals, and remedy weak dimensions of ACWs.

The third stage of this revised methodology would be to utilize the information gathered in the methods described above, conceptualized in Fig. 10.2 below, to finalize and implement a strategy. Figure 10.2 would help identify problems not addressed, effective/missing/problematic instruments, methods of engaging people appropriate for policy problems, in order to tackle moderately structured and unstructured problems and drivers. The strategy would aim at achieving appropriate social learning and would be developed utiliing inclusive development described in Sect. 10.5.

It would also be possible to expand this methodology in depth by utilizing a multi-site, interdisciplinary, multi-team study with a long-term time frame. Adaptive management techniques of experimentation and evaluation could be employed and the full assessment of social learning could be measured over time. The expanded methodology (in time, resources, and depth) would potentially alleviate limitations of ethnocentrism, generalization, and perception bias by employing multiple methods and a larger survey sample. Figure 10.2 is arguably less complex as it combines several steps of the current methodology. Steps one and two are combined into one method of document and content analysis of stage one; Steps four, five and six are combined into a mixed method combination of stage two. Inclusive development, stage three (see Sect. 10.5), is further described below.

10.5 The Need for Inclusive Development

This research answered the question:

How can a theoretical and policy framework (norms, principles, and instruments (regulatory, economic, suasive, and managerial)) be designed to build capacity for rural agricultural producers to respond to the increasing likelihood of d&f, defined by uncertainty?

A theoretical and policy framework to build capacity for rural agricultural producers incorporates adaptive governance. This framework addresses drivers impacting producer livelihoods; it incorporates appropriate missing instruments, effective instruments, instruments enhancing producer livelihood capitals; and it populates areas of problem structuring without instruments (currently unstructured problems of climate change, d&f). Such a framework rejects or fixes ineffective instruments and implements environmental decision making processes of adaptive management, adaptive co-management and anticipatory governance. The most important missing instrument required is one of iterative, inclusive participatory development within governance practices (which has become stage three of the recommended conclusions of this research for the future).

Inclusive development would be inclusive of the anthropocene by helping to cope with the uncertain, systemic, complex problem of climate change, d&f, utilizing the adaptive governance theory and method developed in this research. Inclusive development would also be inclusive in a relational perspective. It would embrace the model of adaptive governance and split ladder of participation and learning (see Sect. 2.4.3) to effectively engage with people. Special attention to meaningful engagement and the inclusion of marginalized people is required. This research has determined that in DCs those without water rights are effectively excluded. The human right of all to water needs to be recognized and implemented. The current concentration of power preventing this has to be disbanded through legitimate participatory processes, changing politics, and embracing social learning. The meaningful inclusion of marginalized people in development and writing the strategy to addresss climate change in the future would be an important step in eliminating poverty and at the same time build resilience to d&f.

Inclusive participatory development aspires to achieve a demonstrated change in understanding through incremental single loop learning (utilizing new procedures, practices or technologies), double loop learning (questioning mental models underpinning strategies and action) and triple loop learning, when values and norms are set aside and a deeper understanding of inter-relationships of people and nature occurs.

References

Argyris C (1999) On organizational learning. Blackwell, Oxford

Bavinck M, Gupta J (eds) (2014) Sustainability science: legal pluralism. Governance and Aquatic Systems, COSUST, 11

Donoso G (2014) Presentation to workshop. From global to local: sustainable water resource governance for Latin America and the Caribbean. At Universidad Catolica de Chile, Santiago Chile, May 5, 2014

Fischhoff B, Watson SR, Hope C (1984) Defining risk. Policy Sci 17:123–139

Gupta J (2014) The history of global climate governance. Cambridge University Press, Cambridge

Gupta J (2016) The Paris Climate Change agreement: China and India. Climate Law 6(1–2):171–181

Gupta J et al (2010) The adaptive capacity wheel; a method to assess the inherent characteristics of institutions to enable the adaptive capacity of society. Environ Sci Pol 13:459–471

Gupta J, Bersgsma E, Termeer CJAM, Biesbroek GR, van den Brink M, Jong P, Klostermann JEM, Meijerink S, Nooteboom S (2015) The adaptive capacity of institutions in the spatial planning, water, agriculture, and nature sectors in the Netherlands. Mitig Adapt Strateg Glob Chang. doi:10.1007/s11027-014-9630-z

Haque E, Burton I (2005) Adaptation options strategies for hazards and vulnerability mitigation: an international perspective. Mitig Adapt Strateg Glob Chang 10:335–353

Hill M (2013) Climate change and water governance. Adaptive capacity in Chile and Switzerland. Springer, Dordrecht/Heidelberg/New York/London

Hurlbert M (2013) Learning, participation, and adaptation: exploring agri-environmental programs. J Environ Plan Manag. doi:101080/096405682013847823. Accessed 15 July 2015

Hurlbert M, Gupta J (2015) The split ladder of participation: a diagnostic, strategic, and evaluation tool to assess when participation is necessary. Environ Sci Pol 50:100–113

Hurlbert M, Mussetta P (2016) Heating resident water governance for irrigated producers in Mendoza, Argentina. Environ Sci Policy 58:83–94

Keen M, Brown VA, Dyball R (2005) Social learning: a new approach to environmental management. In: Keen M, Brown VA, Dyball R (eds) Social learning in environmental management: towards s sustainable future. Earthscan, London, pp 3–21

Leiserowitz A (2007) International public opinion, perception, and understanding of global climate change. UNDP, New York

Moser C (2009) A conceptual and operational framework for pro-poor asset adaptation to urban climate change. Paper presented to the 5th urban research symposium, June 28–30, 2009. Available at: http://siteresources.worldbank.org/INTURBANDEVELOPMENT/Resources/336387-1256566800920/moser.pdf. Accessed 15 June 2015

Obani P, Gupta J (2014) Legal pluralism in the area of human rights: water and sanitation. COSUST 11:63–70

Ostrom E (1998) A behavioural approach to the rational choice theory of collective action. Am Polit Sci Rev 92(1):1–22

Ostrom E (2010) Beyond markets and states: polycentric governance of complex economic systems. Am Econ Rev 100(3):641–672
Pahl-Wostl C (2007) Transition towards adaptive management of water facing climate and global change. Water Resour Manag 21(1):49–62
Pahl-Wostl C (2009) A conceptual framework for analyzing adaptive capacity and multi-level learning processes in resource governance regimes. Glob Environ Chang 18:354–365
Pahl-Wostl C, Giupponi C, Richard K, Binder C, Shervinin A, Sprinz D, Toonen T, van Bers C (2013) Transition towards a new global change science: requirements for methodologies, methods, data and knowledge. Environ Sci Pol 28:36–47
Preston BL, Brooke C, Measham TG, Smith TF, Gorddard R (2009) Igniting change in local government: lessons learned from bushfire vulnerability assessment. Mitig Adapt Strateg Glob Chang 14:251–283
Stirling A (2010) Keep it complex. Nature 468:1029–1031
Reyes B, Salas S, Schwartz E, Espinoza E (2009) Chile governance assessment final report. Available at http://www.parc.ca/mcri/pdfs/papers/gov03.pdf. Accessed 12 Dec 2014
Termeer CJAM, Dewulf A, Breeman G, Stiller SJ (2013) Governance capabilities for dealing wisely with wicked problems. Adm Soc XX(X):1–31
WEF (World Economic Forum) (2013) Global risk 2013, 8th edn. World Economic Forum, Geneva
WWAP (United Nations World Water Assessment Programme) (2015) The United Nations World Water Development Report 2015: water for a sustainable world. UNESCO, Paris
Young OR et al (2005) Science plan. Institutional Dimensions of Global Environmental Change. International Human Dimensions Programme on Global Environmental Change (IHDP), Bonn

Appendices

Appendix I: Interviews

Table A1 Interview details

Study area	Interviews	Pseudonym	Institutions (Examples)
Canada	11	C1–C11	Environment Canada, Ag Canada, Public Safety, Engineers Canada, FCC
Alberta, Canada	8	A1–A8	Environment Canada, Ag Canada, Alta Agriculture, Alta Environment and Sustainable Resource Development
Saskatchewan, Canada	8	S1–S8	Environment Canada, Ag Canada, Water Security Agency, Inter-Governmental Ministry
Coquimbo, Chile	7	Co1–Co7	Inter-American Institute, Conicet, Government Official, Irrigator Association, Water Law Specialists
Mendoza, Argentina	7	M1–M7	Politicians, Water Governance Specialists,
Total	41		

Appendix II: Field Guide for Governance Assessment

A. Purpose and Scope

The Governance Assessment Field Work Guide is an outline of research themes and questions which should be addressed in the semi-structured research interview. ***This is a practical guide, designed to help you with the scope of the interview, and is not designed to be used as a questionnaire – think of it more as a checklist of themes that need to be addressed.*** The goal of the interview is as natural and

M.A. Hurlbert, *Adaptive Governance of Disaster*, Water Governance - Concepts, Methods, and Practice, DOI 10.1007/978-3-319-57801-9

free-flowing a conversation as possible. How you word particular questions, the order they are asked, and how much depth you go into will vary.

The themes outlined below will allow for the collection of information from identified governance organizations and actors which will allow the VACEA project to assess the institutional capital in existence to reduce the vulnerability of the community to climate variability, hazards and extreme events.

B. Set-Up and General Background Preparation

The organizations and people to be selected for interviewing will be chosen based on the institutional profile assembled and the assistance of the VACEA project partners

Prior to the interview, all researchers are expected to have a thorough understanding of the general principles of institutions relating to climate variability, hazards, and extreme events, and water management.

C. Pre-interview

Research the organization – know its mandate and geographic scope, as well as the position of the contact. You should have spent some time with the organization's website and have acquired good knowledge of its areas of concerns, issues, etc. If there are publicly available documents produced by the organization, be familiar with them. (Bottom Line: be thoroughly prepared before you ask people for their time)

D. The Research Interview and Themes

Goal of the Interviews: To ascertain how the institution fits into the governance framework respecting climate variability, hazards, extreme events and water, identify how the organization reduces vulnerabilities of the community, what factors contribute to this, and lastly, whether the characteristics of adaptive governance are present in the organization and its interrelationships with other organizations and community members.

The points below represent **themes** that should be explored. However, the order in which these are addressed and the language used will vary by interview. Depending on the context, there will be instances where particular topics warrant greater depth than the questions below indicate to meet the goal of the interviews. The interview guide assumes that the interviewer has a high degree of familiarity with the subject matter and comfort with open-ended, loosely structured interview techniques.

The main points (in bold) are general themes. The bulleted lists are points you need to address. It is always preferable to gather the information in an open-ended fashion, and you should resort to specific prompts only if necessary. Think of the questions as questions to you, the researcher – you should be able to answer these based on what you learned in the interview?

Part 1: General Information Questions/Setting the Stage

This part of the interview situates the person being interviewed within the larger picture of the institutional profile prepared in the community vulnerability assessment. Details surrounding the entity with whom the person is associated or employed and that capacity should be ascertained. Much of this information should be ascertained from secondary sources and these questions can be only in relation to specific questions the interviewer has after reviewing other sources. Questions might be:

1. **What is the role of the institution with respect to water, climate variability, hazards and extreme events, and what is the role of the respondent within the institution?** (This will assist in assessing the existence of responsiveness and flexibility).
    - What is the role of the institution with respect to climate variability, hazards and extremes, water and climate change? What is its area of institutional responsibility or jurisdiction? How do water and weather condition relate to its mandate?
    - What is the position of the respondent in the organization? What decision-making or administrative tasks relative to water and climate does he/she routinely perform?
2. **What past climate variability, hazards and extremes, or water stress has this organization faced, managed, and mediated, and how?** (This will assist in assessing the existence of responsiveness and flexibility).
    - In what instances has the institution faced climate variability, hazards and extremes, water stress in the past? When? What were the effects of it?
    - Was there an institutional response to climate stress? If so, what was the nature of the response? Was this part of the institution's existing mandate at the time? Did the mandate have to change and if so, how quickly did this occur?
    - In times of crisis, were there unprecedented measures/ad hoc responses which became necessary? How were these implemented? Were new protocols developed? Do early warning systems exist?
    - How flexible has the institution been when it comes to responding to water stress? If the past water stress occurred now, how would things be different?

Part 2: Open-Ended Interviewing on Exposure-Sensitivities, Adaptive Strategies

The purpose of this part of the interview is to ascertain how the entity involved in the institutional profile fits into the vulnerability and adaptive capacity framework. What policies and actions of the entity respond to climate variability, hazards and extreme events? What type of planning occurs?

This part of the interview should be informed by issues that have arisen in the community vulnerability interviews with community member. Based on indications from community members as to their contact with governance organizations in relation to climate variability, hazards and extreme events, the assistance provided by these organizations should be explored or the role of these organizations in relation to the specific climate events identified by the community.

(This will assist in assessing the existence of responsiveness and flexibility).

3. **Does this institution plan for climate variability, hazards and extremes, water/climate stress, and how?**
    - What type of long-term planning is done w.r.t. climate variability, hazards and extremes, water/climate (refer back to routine decisions, past times of stress as needed)? How many years is the planning time frame?
    - How is planning for variability done? What factors are considered? Is there explicit consideration of climate change/long-term scenarios of water availability/moisture deficit/forecasted demand?
    - Are there contingency plans (emergency preparedness or business continuity plans) for particular situations? Are these short, medium, long term? How is the decision made to implement these both procedurally and substantively?
    - How does the institution reassess these plans? Is there monitoring and reevaluation? Is there a set review period? After extreme events are they reviewed? Who has input into the review? Are changes made based on the review?
    - The VACEA field work will have identified various concerns raised by stakeholder during in-community work. Does this organization help address these concerns? Is this part of your official mandate? How does the organization become aware of concerns (link to stakeholder discussion)? How does it know the outcome of actions taken?
    - Does this organization promote capacity building and problems solving in rural communities? How does the respondent define community capacity building?

Part 3: Guided Interviewing

- This portion of the interview provides the basis for assessing adaptive governance within the community of study. The specific aspects of adaptive governance as identified above need to be explored with the interviewee.

4. **What information inputs are used by this institution in its operation and decision-making? How are these obtained? How secure are information flows?** (This will assist in assessing the existence of reflexivity and capacity).
   - What data are routinely used (refer back to points made in other parts of the interview)? What level of information is collected by the institution/individual and what data come from secondary sources? Do scientists work for the institution? Does the institution work with scientists in another institution? How is information obtained from them? How often? In what manner or context?
   - Is local knowledge used by the institution? Does the institution rely just on scientists for information about the ecosystem? Do stakeholders have input into what policy and governance questions are asked?
   - What scientific data modeling does the organization have access to? How does it access this information? What time frame, how frequently?
   - If primary data is collected, what is the purpose of collecting this information (to monitor, to diagnose, to manage)? Does that data permit the identification of problems or early warning? Does the collected information provide the organization with a comprehensive picture of potential problems within its mandate?
   - Is collected data made available to other organizations? To the public? Is this information relevant to rural communities, and if so, is it accessible to them? How do they know about the data collected by the institution, and how is it accessed?
   - Where does the individual/organization get the secondary information he/she needs (agency, contact, informal/formal network of data dissemination)? Are these data public?
   - What data are needed that aren't currently available? What data does the individual/institution have difficulty obtaining?
5. **What resources does the institution have access to, what are its resource constraints, and how does this affect its activities with respect to managing, mediating, and planning for climate variability, hazards and extremes, water-related issues?** (This will assist in assessing the existence of capacity).
   - How is this organization funded? How secure is this funding? What time horizon does funding encompass?
   - Does the organization have the necessary financial and technical resources to carry out its activities? How are further resources sought?
6. **Who are the institution's stakeholders, how do stakeholders relate to the institution, and how is their input incorporated into the institution's management and decision-making?** (This will assist in assessing the existence of equity).
   - Who are the institution's stakeholders? On what basis does the respondent consider them stakeholders? How do the institution and the stakeholders interface? Is there a formal process for soliciting stakeholder input?

- How accessible are decision-makers/planners within the institution to stakeholders? How accessiblt are the institution's scientists to the stakeholders?
- Has the input of stakeholders ever or changed a decision? How are the interests of various stakeholder groups balanced in routine decisions/management activities and times of conflict? Do some stakeholder groups have more influence than others, and why?
- When and how are non-government actors involved in decision making? Do these actors have the opportunity to influence, significantly change, or make major decisions?
- When confronted by a conflict of stakeholders, how does the institution respond? Is the organization sensitive to the various resources available to the different stakeholders? How does access to resources influence/affect significant participation of various stakeholder groups? Is the knowledge base of the different stakeholders considered?
- Does the institution facilitate the process of negotiation of the interests of different stakeholders with respect to particular interests?
- Has the institution's relationship to stakeholders changed over time? How and why?
- Are there areas where the relationship with stakeholders could be better? Why? How?

7. **To whom and how is the institution accountable?** (This will assist in assessing the existence of capacity).

   - Are the governance institutions legitimate or supported by the people? Are accountability procedures in existence?
   - To whom is the institution accountable? What is the process for this (fiscal accountability, progress reports, elections? Against what is this accountability measured (the institution's mandate, public opinion polls, balanced budget…)?
   - How are individuals within the institution accountable (performance reviews from superiors, progress reports)? Does final responsibility rest with any one individual/group of individuals?
   - Are responses to climate variability, hazards, and extreme events equitable to all community members?
   - Are there established ways to monitor/evaluate the success of particular policies/programs? If yes, to whom are the results of this monitoring/evaluation given?
   - What is considered "poor" performance? What are the consequences of this (for the institution, for individuals – decreased funding, less responsibility…)?
   - Have there been changes in the institution in response to poor performance? What happened?
   - Is the institution's performance public record? How is this information made public?

8. **In what networks does this institution operate, and how?** (This will assist in assessing the existence of capacity).

    - To which organizations do you give direction? From which organizations do you receive direction? Which organizations work within parameters heavily influenced by your institution? How does this work? Do these organizations influence how these parameters are set? Is there a formal process for this? How does this work?
    - Which organizations/institutions do you co-manage/collaborate/coordinate with? What is the nature of this collaboration/coordination?
    - Vertical and horizontal networks?